D0221103

DNA Microarrays

Current Applications

Edited by Emanuele de Rinaldis and Armin Lahm

horizon bioscience

Horizon Bioscience
32 Hewitts Lane
Wymondham
Norfolk NR18 0JA
U.K.

www.horizonbioscience.com

British Library Cataloguing-in-Publication Data

A catalogue record for this book is available from the British Library
ISBN: 978-1-904933-25-0

Printed and bound in Great Britain by Cromwell Press Trowbridge

Contents

List of contributors vii
Preface xi

1 Applications for DNA Microarrays: An Introduction 1
Emanuele de Rinaldis

Introduction 1
DNA microarray technologies 2
A typical microarray experiment 7
Applications of microarrays 8
- *Expression profiling 8*
- *Analysis of genomic DNA 12*
Challenges in using DNA microarrays 14
- *Noise 14*
- *Normalization 15*
- *Sample preparation 15*
- *Experimental design 15*
- *Large number of genes 16*
- *Biological factors 16*
- *Array quality assessment 16*
Microarray data analysis 16
- *Unsupervised clustering methods 18*
- *Similarity distances 19*
- *Supervised clustering methods 22*
Public resources for expression data 23
- *Stanford Microarray Database (SMD) 23*
- *Gene Expression Omnibus (GEO) 23*
- *Microarray Gene Expression Data (MGED) 23*
- *ArrayExpress 23*
- *Expression Profiler24*
Book's content 24

2 Gene Networks and Application to Drug Discovery 29

A. Ambesi, M. Bansal, G. Della Gatta, and Diego di Bernardo

Introduction 29
Classifier-based algorithms 31
– *Target identification and validation 32*
– *Hit identification, lead identification, and optimization: mode of action (MOA) 34*
– *Hit identification, lead identification, and optimization: efficacy and toxicity 37*
Gene network reconstruction 39
– *Target validation, hit identification, lead identification, and optimization 45*
Discussion 49

3 Pathway Analysis of Microarray Data 55

Matteo Pellegrini and Shawn Cokus

Introduction 55
Term enrichment analysis 56
Gene set analysis 57
Pathway coherence 59
Reconstruction of networks using expression data 61
Integrated pathway analysis of expression data and transcription factor binding data 65
Conclusions 67
Future developments and trends 67

4 Toxicogenomics: Applications of Genomics Technologies for the Study of Toxicity 71

Uwe Koch

Introduction 71
Toxicogenomics in the preclinical testing funnel 73
Mechanistic toxicogenomics 73
Phenotypic anchoring 74
Examples 76
– *Proteomics 76*
– *Metabonomics 77*
Predictive toxicogenomics 78
– *Example 78*
Data analysis 79
In vitro methods 80
– *Example 80*
Absorption 80
Genotoxicity 81
Integration 81
– *Computational tools 81*
– *Data 82*

5 Microarray Gene Expression Atlases 89

John C. Castle, Chris J. Roberts, Chun Cheng, and Jason M. Johnson

Introduction 89
Primary uses 90
- *Where is my gene expressed? 90*
- *What genes are expressed in my tissue? 91*
- *What groups of gene are specific to my tissue? 91*
- *Biomarkers: which genes are expressed specifically in condition X? 94*

Additional uses 95
- *Tissue correlation comparisons 95*
- *Gene correlation comparisons 95*
- *Species correlation comparisons 97*
- *Transcription factors and targets 97*

Conclusion 99

6 Supervised Classification of Genes and Biological Samples 101

Adrian Tkacz, Leszek Rychlewski, Paolo Uva, and Dariusz Plewczynski

Introduction 101
Nearest shrunken centroids (NSCs) 103
k-Nearest neighbor 104
Support vector machine (SVM) 107
Supervised neural networks 109
Random forests (RFs) 111
Conclusions 112
Acknowledgements 112
Appendix A: Classification algorithms 112
- *Support vector machine (SVM) 112*
- *Nearest shrunken centroids (NSCs) 113*
- *k-Nearest neighbor (kNN) 114*

Appendix B: Analysis software systems 116
- *Gene Cluster 2.0 116*
- *PAM 116*
- *Genesis 117*
- *Stuttgart Neural Network Simulator (SNNS) 117*

7 A Case Study: The Mammary Carcinogenesis in HER2 Transgenic Mice 121

Federica Cavallo, Guido Forni, Anna Grassi, PierLuigi Lollini, and Raffaele Calogero

Introduction 121
- *The HER2 oncogene 122*
- *The BALB-neuT breast carcinoma murine model 122*
- *Halting BALB-neuT carcinoma by HER2 vaccination 123*

Vaccination effects and transcriptional profiling 124
- *Triplex vaccination 124*
- *Prime-and-boost vaccination 126*

Deriving tumor-associated antigens (TAAs) from BALB-neuT breast carcinoma transcription profiles 127
Bioinformatic approaches for microarray data analysis 129
- *Oligonucleotide chips 129*
- *GeneChips data analysis 130*

Conclusions 137

8 DNA Microarrays: Beyond mRNA 143

Armin Lahm

Introduction 143
mRNA variants (splicing) 144
Genomic tiling: transcription outside annotated regions of the genome 146
Profiling of non-coding RNAs 148
DNA methylation 149
"ChIP-on-chip" for mapping protein–DNA interactions and epigenetic marks 152
Detection of genome variations 155
Integrative genomics 156
Microarrays focusing on viruses and bacteria 156
Conclusions 157

Index 169

Contributors

A. Ambesi
Telethon Institute of Genetics and Medicine
Naples
Italy

M. Bansal
Telethon Institute of Genetics and Medicine
Naples
Italy

Diego di Bernardo
Telethon Institute of Genetics and Medicine
Naples
Italy

dibernardo@tigem.it

Raffaele Calogero
Department of Clinical and Biological Sciences
University of Turin
Orbassano
Italy

raffaele.calogero@unito.it

John C. Castle
Rosetta Inpharmatics LLC
Merck & Co.
Seattle, WA
USA

john_castle@merck.com

Federica Cavallo
Department of Clinical and Biological Sciences
University of Turin
Orbassano
Italy

Chun Cheng
Rosetta Inpharmatics LLC
Merck & Co.
Seattle, WA
USA

Shawn Cokus
University of California
Los Angeles, CA
USA

G. Della Gatta
Telethon Institute of Genetics and Medicine
Naples
Italy

Guido Forni
Department of Clinical and Biological Sciences
University of Turin
Orbassano
Italy

Anna Grassi
Bioinformatics and Genomic Group
CNR - Institute of Biomedical Technologies Section of Bari
Bari
Italy

Jason M. Johnson
Rosetta Inpharmatics LLC
Merck & Co.
Seattle, WA
USA

Uwe Koch
Istituto Di Ricerche Di Biologia Molecolare P. Angeletti
Merck Research Laboratories Rome
Pomezia
Italy

uwe_koch@merck.com

Armin Lahm
Bioinformatics Group
Istituto Di Ricerche Di Biologia Molecolare P. Angeletti
Merck Research Laboratories Rome
Pomezia
Italy

armin_lahm@merck.com

PierLuigi Lollini
Cancer Research Section
Department of Experimental Pathology
University of Bologna
Bologna
Italy

Matteo Pellegrini
University of California
Los Angeles, CA
USA

matteop@mcdb.ucla.edu

Dariusz Plewczynski
Interdisciplinary Centre for Mathematical and Computational Modelling
University of Warsaw
Warsaw
Poland

D.Plewczynski@icm.edu.pl

Emanuele de Rinaldis
Bioinformatics Group
Istituto Di Ricerche Di Biologia Molecolare P. Angeletti
Merck Research Laboratories Rome
Pomezia
Italy

emanuele_derinaldis@merck.com

Chris J. Roberts
Rosetta Inpharmatics LLC
Merck & Co.
Seattle, WA
USA

Leszek Rychlewski
BioInfoBank Institute
Poznan
Poland

Adrian Tkacz
Bioinformatics Unit
Department of Physics
Adam Mickiewicz University
Poznan
Poland

Paolo Uva
Bioinformatics Group
Istituto Di Ricerche Di Biologia
Molecolare P. Angeletti
Merck Research Laboratories Rome
Pomezia
Rome
Italy

Preface

Much has been written on the perspectives opened up by the completion of the human genome to the biomedical sciences. The knowledge of the complete repertoire of genes contained in human genetic material allows for a conceptual scale-up in areas such as biochemistry, molecular and cell biology.

These disciplines are progressively moving from a traditional single-gene/protein focus to a more complex, integrated, and dynamic view of the molecular events taking place in the cell. As a result, the attention to individual entities such as genes, proteins, and gene mutations is progressively shifting toward the analysis of functional gene networks, biochemical pathways, and complex genetic features such as haplotypes.

DNA microarrays is the technology that probably has contributed most to the consolidation of the "global" approach to biological experimentation that started with the human genome era. Given its potential in areas such as pharmacogenomics, diagnostics, and drug target identification, this technology has polarized the enthusiasm and the expectations of the scientific community and has been the hub of important investments by industrial and academic institutions. Furthermore, the advent of microarrays has given a strong impetus to computational biology through the need for new tools and algorithms for data analysis and has shortened the distance between areas such as molecular biology, clinical diagnostics, statistics, information technology, and epidemiology.

At present, it seems clear that large-scale microarray studies are not a passing fashion, but are instead becoming a crucial aspect of a new way to conceive experimental biology, one involving large-scale, high-throughput assays.

With this book, we aim to outline the major results achieved since the first appearance of DNA microarrays in the scientific scene. Our intention is not only to illustrate a number of successful applications but also to provide the reader with useful guidance of the main concepts and the philosophy driving these types of study. By integrating the description of the methodologies with the scientific achievements obtained in various research areas, we hope to highlight the potential and the challenges offered by high-throughput approaches to modern biology.

Finally, and maybe more importantly, we hope that readers will find this field as interesting as we do.

Emanuele de Rinaldis

Applications for DNA Microarrays: An Introduction

1

Emanuele de Rinaldis

Abstract

DNA microarrays are the modern, parallel version of the classic molecular biology techniques of Northern and Southern blotting. While the blotting techniques are capable of detecting the abundance of a specific nucleotide sequence in a biological sample, DNA microarrays allow the exploration of thousands of sequences in a single run. The difference in data throughput has profound implications in the nature of the information that can be derived, as it allows complete analysis of the genetic material and the monitoring of virtually all the nucleotide sequence molecules occurring in a biological sample under various conditions. This allows the traditional reductionist view to be replaced by a more complex and integrated view of the molecular events taking place in the cell. As a consequence, analytical methods of investigation based on computational/statistical techniques are required for the interpretation of the high volume of data generated.

In this chapter a general introduction on DNA microarrays is given, illustrating the existing technologies, the fields of application and the principal aspects and issues related to the production and the interpretation of microarray data.

Introduction

Like many techniques widely used in molecular biology, the fundamental basis of DNA microarrays is the process of hybridization. Two strands of nucleic acid, DNA or RNA, hybridize if they are complementary to each other. This principle is exploited to measure the unknown quantity of one RNA or DNA molecule (target) on the basis of the amount of a complementary sequence (probe) that has hybridized to the target. The level of hybridization is quantified by measuring the level of a detectable chemical label, used to mark the target or the probe sequence in the experiment.

In the microarray technique, the probe sequences are immobilized on the surface, at a separation of a few micrometers, so that it is possible to place many different probes on a small single surface of one square centimeter. The sample is usually labeled with a fluorescent dye that can be detected by a light scanner that scans the surface of the chip. Each probe matches a particular nucleotide sequence present in the sample material, a preparation of RNA or DNA. After the hybridization of the sample material on the chip, the signal detected on the spots upon light scanning, measures the abundance of each specific nucleotide sequence present in the sample. Observing all the microarray spots at the same time gives the *profile* of a sample; in the most common application of DNA microarrays, the target sample is mRNA and the total microarray image represents the *transcriptional profile* of the sample.

Microarrays are therefore a means of spatially arranging molecular probes, so that their specific signals can be independently estimated and the potential applications can include analysis and characterization of any reaction product composed of nucleotide sequences.

For example, besides the widely used applications of microarrays to gene expression, microarrays can be used for genotyping analysis (Hacia *et al.*, 1999) or detection of splice variants (Shoemaker, 2001). Gene expression profiling can also be studied by using *proteomics* technologies, devoted to the high-throughput study and characterization of the entire collection of proteins contained in a biological sample.

As proteins are the real functional effectors in the cell, this approach represents a more direct means of monitoring cell functions. However, in some cases mRNA levels have revealed to be even more informative than protein levels as they can show adjustments of the cell to individual protein functional disruptions (Hughes *et al.*, 2000), even if protein levels are unchanged. Thus, the transcriptional expression profile of a certain biological sample can be considered, to a great extent, a precise functional indicator of a specific biological state.

Another important aspect in favor of DNA microarrays is the technological advantage in terms of resolution. While the total number of mRNA messages transcribed by a genome (*transcriptome*) can be monitored by standard chips, the monitoring of the corresponding proteins (*proteome*) still represents a challenge for proteomics technologies.

DNA microarray technologies

DNA microarray technologies can be classified according to the method used for the deposition of the probe sequences—oligos or cDNA—on the chip, that can be done by *presynthesis* or *in situ* synthesis. Tables 1.1 and 1.2 summarize the different technologies available for the two categories, some of their main characteristics and the vendors (Stoughton, 2005).

The presynthesis array technology is often adopted by small laboratories that want to have the freedom to target specific, often non standard, sequences and therefore need to design their own chips with ad hoc probes. Arrays based on presynthesis of the probes (spotted arrays) allow more flexibility in the selection of desired sequences and, what can be essential for particular experimental designs, the DNA of unknown sequences can also be spotted. In presynthesis

Table 1.1 Microarray alternatives using presynthesized oligos or cDNA (Stoughton, 2005)

Process	Vendors	Substrate	Density	Advantages	Limitations
Pen tip deposition	—	—	—	Low technology investment	Deposition not highly repeatable, droplet size limits density
	Clontech Pharmingen Sigma-Genosys	Nylon or other synthetic polymer	<10 features per mm^2	Low manufacturing cost per array	Substrate properties or pen tip diameter limit feature density
	Clinical MicroSensors	Printed circuit board	<10 features per mm^2	Potentially low manufacturing cost per array	Insensitive detection, substrate, chemistry difficult to control
	Clontech Pharmingen Harvard Biosciences Mergen MWG Biotech	Glass	<100 features per mm^2	Potentially very low manufacturing cost per array	Diameter of pen tip and droplet size limit density
Ink-jet deposition	GE Healthcare (CodeLink arrays)	Polyacrylamide gel on glass	100 features per mm^2	Potentially very low manufacturing cost per array	
Electro-phoretically driven deposition	Nanogen	Silicon (semiconductor fabrication process)	100 features per mm^2	Fast hybridization, relatively easy to produce custom arrays	Expensive chip fabrication process, electrophoretic process not fully developed

Table 1.2 Microarray alternatives that use *in situ* synthesis (Stoughton, 2005)

Process	Vendors	Substrate	Density	Advantages	Limitations
Light-directed synthesis	—	—	—	Potential densities approach wavelength of light	Photo-deprotection stepwise yields limit oligo length
Photolithographic mask for each of the four bases at each layer	Affymetrix	Glass	8200 features (360 transcripts) per mm2	Leader in density and quality control	Manufacturing costs of photolithographic masks
Digitally controlled micromirrors incorporated in chip	NimbleGen Systems	Glass	1000 features per mm2	Easy to produce small lots of custom arrays	
Digitally controlled micromirrors and microfluidics incorporated in chip	FeBit, Xeotron	Glass/silicon (semiconductor fabrication process)	1000 features per mm2	Easy to produce small lots of custom arrays	Manufacturing costs of microfluidic structures
Ink-jet printer head deposition of nucleotides layer by layer	Agilent Technologies	Glass	100 features per mm2	Easy to produce small lots of custom arrays	Feature density limited by feasible droplet sizes
Electrode-directed synthesis	CombiMatrix, Nanogen	Silicon (semiconductor fabrication process)	<100 features per mm2	Custom arrays	Fabrication process still expensive

based microarrays a robot spotter is used to place small quantities of probe in solution from a microtiter plate to the surface of the chip (Fig. 1.1a).

In arrays based on the *in situ* synthesis the steps of cloning, amplification by PCR and probes spotting are skipped, with the important advantage of reducing the noise and the variability of the system.

Most array vendors offer standard, ready-to-use chips designed for monitoring the expression profiles or the genotype of many common organisms, like human, mouse, rat, yeast. In this case the customer is relieved from the complex problems of designing probes and placing them on the chip.

The Affymetrix technology uses the photolithographic method to synthesize probes of oligonucleotides *in situ*. On Affymetrix chips target genes are represented by a set of probe pairs (about 16–20 per gene, depending on the array), each composed of an oligonucleotide (usually a 25-mer) perfectly matching the target sequence (PM probe) and an oligonucleotide containing a mismatch (MM probe). The set of probe pairs for a target sequence is called "probe set" (Fig. 1.2). This strategy allows an estimate of the signal caused by non specific target–probe interactions.

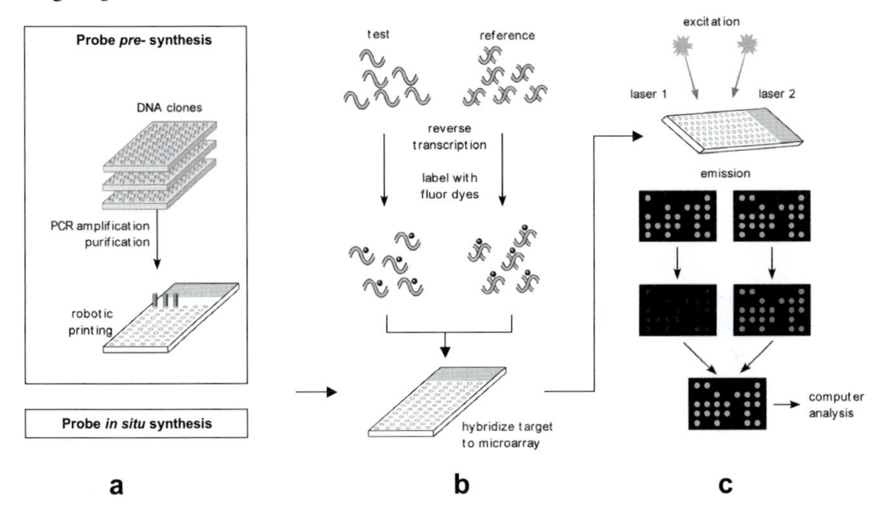

Figure 1.1 Gene expression profiling experiment on a double channel microarray. (a) Probe sequences are deposited on the chip by presynthesis or *in situ* synthesis. (b) mRNA from a test and a reference sample is extracted, reverse transcribed and amplified. Labeling molecules are incorporated during the synthesis of the amplification products. The two labeled samples are then hybridized on the chip. (c) The chip is excited at different wavelengths, one for each of the fluorophores used and the fluorescence intensities of the spots are measured. Red fluorescence of each spot measures mRNAs abundances in one sample (e.g., test) and the green fluorescence measures the same mRNAs in the other sample (e.g., reference). The two images can be merged: red spots and green spots in the combined image indicate respectively a prevalence of probe hybridization in the test sample (gene upregulation), or in the reference sample (gene downregulation). Yellow spots in the combined image indicate equally expressed genes in the test and reference samples. This figure is also reproduced in color in the color section at the end of the book.

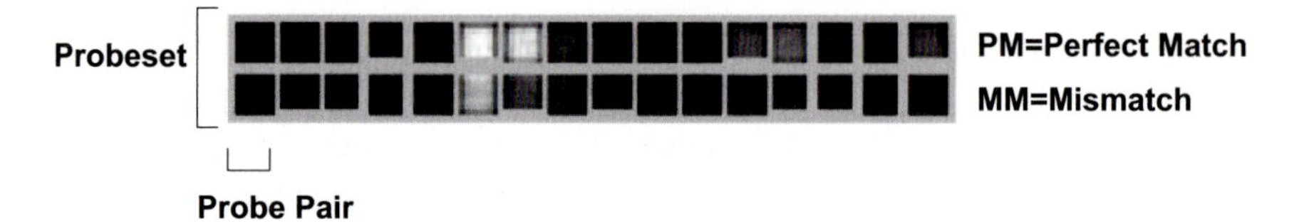

Figure 1.2 In Affymetrix, chips target genes are represented by a set of probe pairs (16 to 20 for gene, depending on the array), each composed of an oligonucleotide (usually a 25-mer) perfectly matching the target sequence (PM probe) and an oligonucleotide containing a mismatch (MM probe). The set of probe pairs for a target sequence is called "probeset."

Another leader *in situ* technology is the ink-jet, adopted for example by Agilent. This has the advantage of being easily customizable and allows longer oligonucleotides to be placed on the chip (60-mer). This aspect is important because, as shown by a recent publication (Hughes *et al.*, 2001), 60-mers have better sensitivity-specificity trade-off than shorter oligos. The main limitation, compared with the Affymetrix technology, resides in the lower probe density (feature density) of the chip. As a consequence, each gene can be represented by no more than one or a few probes.

Another relevant aspect which distinguishes array technologies is whether they are "single" or "multi" channel. Affymetrix arrays are single channel, meaning that only one sample can be assayed per chip. Double-channel arrays (Agilent, Clontech and others) allow instead a two-color competitive hybridization of two samples on the same chip. Two samples can be therefore labeled with different fluorophores, mixed together and hybridized to the chip. Upon laser scanning at different wavelengths, the signals relative to the two samples can be measured independently. As will be illustrated in the next section, this approach can be used to directly generate expression ratios between a test and a reference sample (e.g., tumor vs. normal).

The advantage here is that, being the two samples assayed on the same chip, the biases due to the external sources of variation (chip quality, hybridization conditions) are intrinsically reduced. Another favorable aspect of multichannel arrays is that the number of chips needed for the experiments can be reduced as more mRNA samples can be gathered using just one chip.

However, there are also well-known drawbacks associated to multichannel arrays:

1 The design of large experiments involving a large number of mRNA samples can become more complex than with single channel arrays.
2 Some genes may incorporate certain dyes with different efficiency causing gene-specific dye biases.
3 Dyes can have global different efficiencies.

A typical microarray experiment

To illustrate the general workflow of a microarray experiment, we describe here an example in which DNA microarrays are used to compare the expression profiles of a tumor and a normal biological sample. We assume the use of commercial double-channel arrays, containing probes for many thousands of human genes (Fig. 1.1a).

The protocol begins with the extraction of the mRNA from both samples, followed by an amplification/labeling step, normally obtained by some combination of reverse-transcriptase/PCR/*in vitro* transcription; the final product can be either cDNA or cRNA. Labeling molecules are incorporated during the synthesis of the amplification products (Fig. 1.1b).

As a starting point we consider the fluorophores Cy5 and Cy3 (respectively represented by the pseudocolors red and green) to be used to label respectively the tumor and the normal sample. The two samples get co-hybridized on the chip so that the target sequences present in the two samples can anneal with their specific probes placed on the array. In general, the hybridization step lasts several hours and occurs in highly stringent conditions of buffer and temperature to increase the annealing specificity. A washing phase follows, to remove any excess of the labeled samples and the chip gets dried.

The final experimental step, before the downstream "in silico" analysis of the data, is the laser scanning of the array. In this phase, the chip is excited at different wavelengths, one for each of the fluorophores used, and the fluorescence intensities of the spots are measured. As a result in our example we would have the red fluorescence of each spot measuring mRNAs abundances in the tumor sample and the green fluorescence measuring the same mRNAs in the normal sample. Each scan produces indeed a separate fluorescence image of the array (Fig. 1.1c). In the absence of particular biases due to dye efficiency, the merging of the two red and green images allows a preliminary, visual identification of: (1) genes prevalently expressed in the tumor sample ("upregulated" in tumor, red spots); (2) genes prevalently expressed in the normal sample ("downregulated" in tumor, green spots); (3) genes not detected in any of the two samples (black spots); and (4) genes equally expressed in tumor and normal sample (yellow spots). After data normalization and transformation procedures, the gene intensity values in the two samples can be reported in a scatterplot. Genes equally expressed in the two samples fall around the diagonal, whereas differentially expressed genes, up or downregulated, can be identified as "outlier" points in the plot. For simplicity we have described a single microarray experiment. However, it is common practice to have at least two technical replicates for each of the biological sample assayed. In the case of double-channel arrays, this is usually achieved by coupling each experiment with a relative "dye swap" experiment, in which the same samples are labeled in reverse order. In our example, Cy3 and Cy5 would then be used to label the tumor and the normal samples, respectively, and both samples would get assayed on a second array.

The two scans obtained for each sample by using the two fluorophores, on two arrays, are eventually averaged. This way the variations due to chip specificities or to different label efficiencies get mostly mitigated.

Applications of microarrays

Microarrays have been used successfully in a variety of applications, which include sequencing (Schena *et al.*, 2000), single nucleotide polymorphism (SNP) detection (Wang *et al.*, 1998), characterization of protein–DNA interactions (Rodriguez *et al.*, 2005), DNA computing (Kary *et al.*, 1997), and others. However, mRNA profiling applications currently dominate microarray usage because of the rich information that can be derived about the functions of genes in cells and tissues. These data can become even more useful when combined with additional data types from the same biological samples (e.g., anatomical and clinical data in disease studies, proteomics data and phenotypes). A good review of the applications of DNA microarrays in biology can also be found in Stoughton *et al.* (2005).

Below is reported a selection of typical biological questions that can be addressed by using DNA microarray. Although they are described separately, many of them are closely related and the distinctions are mainly due to illustration purposes.

Applications can be divided into two main classes: mRNA gene expression profiling and analysis of genomic DNA. A large part of the examples mentioned here will be more extensively illustrated in the next chapters.

Expression profiling

Identify genes whose expression depends on a specific biological state (e.g., a disease)

- Which are the genes affected by a certain condition?
- What are the genes' determinants for a certain cell state?

What comes out of these types of experiments is normally a set of values corresponding to the relative abundance of gene mRNAs in two biological conditions, such as a disease state versus a normal state. Genes differentially expressed in the disease state can be identified and, especially if other in silico or experimental observations are available, hypotheses can be made to identify the genes playing a causal role in the development of the disease. These genes, if the transcriptional data are also confirmed at protein level, can have a potential interest as candidates for drug targets in pharmaceutical research. Ideally, drugs can be designed to specifically inhibit any particular gene, protein or signaling cascade and, if the target is specifically expressed in the diseased tissue, there is less chance of causing undesirable effects.

Hundreds of scientific papers are currently available reporting lists of genes differentially expressed in various biological samples and under various conditions. An overview of the wealth of available data can be obtained by exploring public data repositories such as GEO or Array Express (see below).

Monitoring transcriptional response to variable conditions

- How global gene expression is remodeled during development, or upon drug treatment?
- What genes play a central role in the cellular response to an external stimulus?

These experiments explore the expression profiles upon systematic variations of set of conditions, like time, cell development, drug treatment. Progression of gene expression during development was for example monitored in *Drosophila* (White *et al.*, 1999) and in *Caenorhabditis elegans* (Kim *et al.*, 2003), in which genes could be grouped according to their pattern of expression over different phases of development. These co-regulation based groupings are a fairly accepted way in microarray data analysis to gain functional information using the "guilt by association" inference (Quackenbush, 2003): a gene of unknown function is predicted to be associated to a functional role if its expression pattern is similar to that of a functionally known gene. One of the first and more convincing examples of this type of studies is the work of Hughes *et al.* (2000) where a large set of different single-gene disruption mutants in yeast were compared by transcriptional analysis. The resulting patterns could be clearly visualized in two-dimensional heatmaps (see Fig. 1.4, below) and provided functional inferences for previously unannotated genes.

Studying pathways and biological gene networks

- What are the pathways perturbed in particular conditions?
- How genes influence each other?

A method of analyzing microarray data that is becoming standard is the analysis of pathways. While standard gene expression analysis looks at each gene as an independent entity, pathways analysis is aimed at the identification of coordinated changes in expression, affecting many genes at the same time.

The idea at the basis of the various approaches and tools for pathway analysis is that of characterizing the behavior of groups of genes (often referred as "modules" or "gene sets") that act in concert to carry out a specific function. Defined gene sets can include annotation-based groups of genes belonging to the same functional category/pathway or can derive from the analysis of independent expression data, like clusters of co-expressed genes, genes expressed in particular tissue types or in particular conditions. Additional types of annotations could also be used to group genes, such as promoter elements, chromosome position, protein structure or interaction data, and text enrichment in the literature. The pre-classification of genes into groups on the basis of different criteria and the observation of their expression behavior, allows looking at the data from different perspectives, like for example examining what happens to the expression of the membrane protein genes or to the kinases, as a whole, in particular conditions. The detection of small, but consistent, changes in expression of a group of genes with related function, allows the elucidation of biological aspects and trends that would be hidden in a classic gene-by-gene analysis.

An example is represented by the work of Mootha *et al.* (2003) were the expression of sets of genes, grouped according to pathways, were tested for association with disease phenotypes.

The method was applied to a microarray dataset of human diabetic muscle and they showed that by examining the joint behavior of the gene sets, significant changes could be detected even in cases where the expression of individual genes was not significantly different. Molecular processes systematically altered in the diabetic muscle could be revealed only in the coherent signal associated with the higher-level, gene sets entities.

Another approach to glean "global" information from array data is the reconstruction of functional, regulatory networks from gene expression, by "reverse engineering" techniques. These techniques are aimed at characterizing the relationships among genes (e.g., A activates B, which represses C) starting from their expression values in various conditions (e.g., time series or gene knockouts). Most of these methods are tested on previously published data, from which they rediscover some known relationships, propose revisions or contradictions of others, and suggest many novel interactions.

An area in which this approach has shown to be very promising is the study of the drug's molecular mechanism of action (MOA). De Bernardo and collaborators have presented in a recent work (Di Bernardo *et al.*, 2005), also illustrated in the next chapter, a machine learning method for the analysis of gene expression data in response to drug treatment. The method reconstructs regulatory networks from transcriptional data and is able to discriminate between direct effects of the drug on gene expression from downstream, secondary effects. From the generated functional genes network, the gene or the biochemical pathways directly targeted by the drug can eventually be inferred.

Another common way to take advantage of expression profiles data is to use them as a *fingerprint* of a given biological state. The idea is to exploit the analysis of many hybridization experiments to identify common patterns of gene expression among samples showing similar biological characteristics.

This approach is not mainly directed towards the understanding of the biology behind a given phenotype but is rather aimed at finding associations between gene expression behavior and a phenotype. This information can be used for classifying new samples according to their expression profiles.

The next two sections illustrate some common applications of this approach.

Disease diagnosis, prognosis, and treatment

- Is there an expression profile pattern that can be used to infer diagnostic/prognostic information?
- Can the optimal drug treatment be associated to a particular pattern?

One of the most promising application of microarray in biomedical research is the identification of candidate genes to be used as biomarkers of a particular phenotype.

In this respect, even groups of genes whose perturbation does not have a causal role in the origin of a disease could be of interest for their diagnostic potential. In general, the detailed molecular phenotype provided by expression profiling has a level of resolution much higher than conventional classification methods (e.g., microscopy analysis of biopsies). This allows a more subtle discrimination between states showing the same gross phenotype but for which the subsequent progression of molecular events differs.

Eventually, more precise and less invasive clinical diagnosis procedures could be developed (Mischel *et al.*, 2004). The accurate classification of different disease subtypes, is of crucial importance also for decisions regarding drug treatment as different subtypes are likely to require different treatments. The scientific discipline aimed at studying relationships between genomics (transcriptional and/or genotyping data) and pharmacology (e.g., the response to a drug treatment) takes the name of *pharmacogenomics*. The idea of a "personalized medicine," tailored on the basis of patient specific expression signatures and/or genotype, has strongly contributed to the growth of microarray based studies and represents one of the most important promises of the genomics era.

An example of profiling data used as diagnostic/prognostic tool, is the famous work on acute lymphoblastic leukemia and acute myeloid leukemia (ALL and AML, for short) published by Golub *et al.* (Ramaswamy *et al.*, 2001). There is a close relationship in these studies between the recognition of sub-types of a disease for diagnostic purposes and the development of prognostic predictors. In fact, by exploiting statistical associations between clinical follow-up information of patients' cohorts and the occurrence of specific expression patterns, microarray experiments can be used to derive prognostic information. Golub *et al.* have used microarray data for the construction of specific disease pattern profiles and demonstrated, by cross-validation methods, the power of using profiles as diagnostic and prognostic tools. Cancer outcome prediction studies are very popular and many published evidences suggest that in the near future microarray could be employed as routinely prediction tool for cancer prognosis.

Alizadeh *et al.* (2000) have shown that a specific expression pattern is indicative of survival in B-cell lymphoma patients and characteristics of two subtypes of large diffuse lymphoma B-cells. Another example is represented by the work of Van't Veer *et al.* (2000), in which the authors were able to find an algorithm which combines the expression levels of the 70 identified transcripts to predict metastasis of breast tumors.

Prediction of pharmacokinetics/toxicity and drug screening

- Given the expression profile of a sample treated with a compound, can the pharmacokinetics/toxicity behavior of that compound be inferred?

Microarrays have also been used for the assessment of the toxicity based on changes in gene expression. The idea behind can be schematically illustrated as follows: a collection of signatures associated with toxicity can be built by exposing cells or tissues to different classes of chemicals whose toxicity is known.

Then, when the same cells or tissues are exposed to a chemical with unknown toxicity, the expression profile results could be compared with the pre-built collection of toxicity related signatures. The degree of profile similarity can be used to infer the toxicity properties of the compound. A similar approach can be applied to predict the pharmacokinetic behavior. After the statistical validation of the approach, the process could be automated to allow for high-throughput toxicity/pharmacokinetics screening of new molecular entities and can reduce the need for lengthy, expansive and unpleasant testing of potential drugs.

A comprehensive review of this type of studies can be found in Chapter 4.

Analysis of genomic DNA

Check changes in gene copy numbers

- Are genes amplified at DNA level?
- Are amplifications associated to a disease?

Change in gene copy number is another mechanism by which gene expression can be altered. DNA microarrays can be used to directly measure the concentration of genomic DNA fragments from particular genomic regions. Lucito et al. (2003) have used microarrays in this way to scan changes in the gene copy number associated to cancer. Another example is the application of DNA microarray for the study of the temporal progression of replication along the chromosome (Raghuraman et al., 2001).

Identify transcription factor binding sites

- Where does a given transcription factor bind DNA?

Another goal of genomic studies is the identification of the complement of regulatory DNA sequences that are bound by transcriptional regulators. DNA microarrays can be used for a genome-wide identification of *in vivo* transcription factor binding sites by chromatin immunoprecipitation (ChIP) coupled with array hybridization (ChIP–chip) (Buck *et al.*, 2004). The possibility of using overlapping probes on the array with complete genomic coverage allows the potential identification of all DNA binding regions for a specific transcription factor. Indeed, at a high probe tiling resolution, multiple overlapping probes may contain the actual transcription factor binding motif and thus enable a fine mapping of the binding site to a resolution of less than 25 bp. Successful examples of the application of the ChIP–chip approach on yeast and mammalian cells can be found respectively in Iyer *et al.* (2001) and Martone *et al.* (2003). In the first study the binding sites for two transcription factors in yeast were identified whereas the second allowed the identification of NF-κB binding sites across human chromosome 22.

Characterization of microbial pathogens

- Are pathogens present in a sample (air, water, organ tissue)?
- What are genetic differences between two microbial strains?

The possibility to use DNA microarrays to characterize the presence of specific genetic sequences can be exploited for parallel interrogation of pathogen genomes. The presence of pathogens causing infective diseases can therefore be monitored in biological samples, food, water, air, etc.

Moreover, complete genomes can be compared by using microarray probes specific for genes present in a given microbial strain, which is used as baseline. DNA from other strains can be directly compared by competitive hybridization with the baseline strain and the degree of genetic similarity between the target and the baseline pathogen strain can be assessed. This approach was successfully applied in the identification of the differences between *Mycobacterium tuberculosis* and the associated bacillus Calmette–Guerin vaccine strain (Behr *et al.*, 1999) and to show near genetic identity between strains responsible for two separate epidemics of rheumatic fever caused by group A *Streptococcus* (Smoot *et al.*, 2002).

Genotyping

- What are the genetic mutations in a genome?
- Can some of them be associated to a particular phenotype?

DNA microarrays can be designed for the genome-wide identification of single nucleotide polymorphisms (SNPs). The large usage of microarray for this application has greatly contributed to the huge volume of human SNPs data currently available as part of the human genome project.

When the interest is focused on specific lists of known polymorphisms scattered throughout the genome, probe sets can be designed just for them and chips can be used as a tool for rapid screening of biological samples. Chips have been indeed designed to detect mutations in genes of particular interest to human health such as the cystic fibrosis gene CFTR (Cronin *et al.*, 1996), the breast cancer susceptibility gene BRCA1, P53 (Ahrendt *et al.*, 1999), and mitochondrial DNA (Stoughton, 2005). Mutations associated with drug response can also be detected. In a recent work Ahmadi *et al.* (2005) have identified a set of SNPs that represents the common variations in genes controlling the absorption, distribution, metabolism and excretion of drugs and demonstrated their utility for pharmacogenetics research. Performance studies in the context of P53 (Wikman *et al.*, 2000), and mitochondrial DNA (Cutler *et al.*, 2001), have shown the great opportunities offered by this approach as a tool for diagnostic and/or prognostic, together with the limitations associated with false detections when the underlying mutation rates are low.

Challenges in using DNA microarrays

As we have seen, DNA microarrays have established themselves as a leading technology in a large range of applications. However, there are still a number of aspects which represent a challenge for scientists working in the field and stimulate a continuous research of experimental and computational solutions. It is out of the scope of this chapter to go into the details of all of them; nonetheless it is worthwhile to have a general idea of the main issues related to the production and the analysis of microarray data.

Noise

Due to the complex sequence of events underlying a microarray experiment, the resulting data tend to be very noisy. Noise can be introduced at virtually each step of the procedure: mRNA preparation and amplification, labeling, probe deposition, chip surface chemistry, humidity, target volume, chip inhomogeneities, target fixation, hybridization parameters, unspecific hybridization, non-specific background hybridization, artifacts, scanning, image segmentation, quantification etc.

The resulting noise represents a strong confounding factor when two different biological samples are compared and it is not an easy task to discriminate whether the observed differences are due to artifacts or to real biological changes in gene expression.

The only way to minimize the technical noise is to augment the number of replicated experiments. It is also very important to invest time and resources in the preliminary set-up of experiments, aimed at identifying the main sources of technical variation and providing a realistic estimate of the sensibility and sensitivity limits of the technology. A list of some of the main noise sources is shown in Table 1.3.

Table 1.3 Sources of fluctuations in microarray experiments (Draghici, 2003)

Source	Description
mRNA preparation	protocols and kits can vary
Transcription	depends on th type of enzyme used
Labeling	depends on the label type and age
Amplification (PCR protocol)	quantitative differences in different runs
Pin geometry variations	different surfaces due to chip production errors
Target volume	fluctuates stochastically
Target fixation	fraction of target cDNA linked to the chip's surface unknown
Hybridization parameters	influenced by many factorz such as temperature, time, buffering and others
Slide inhomogeneities	batch to batch variations
Non-specific hybridization	hibridization to background or to not complementary sequences
Gain setting (PMT)	shifts the distribution of the pixel intensities
Dynamic range limitations	Variability at low end or saturation at the high end
Image alignement	Not aligned images of the same array at various wavelengths
Grid placement	Center of the spot not located properly
Non-specific background	Erroneous elevation of the average intensity of the background
Spot shape	Irregular spots are hard to segment from background
Segmentation	Bright contaminants can seem like signal
Spot quantification	Pixel mean, median

Normalization

Normalization is aimed at minimizing identified technical biases like different dye efficiencies, different quantities of loaded material on the arrays, dye non-linearity and saturation towards the extremities of the range. Several normalization procedures are available. Choosing the most suitable algorithm for a given dataset and verifying the assumptions on which the algorithm is based (e.g., normal distribution of gene expression values or similar number of up and down regulated genes in a two-sample comparison) can be a not easy task. A good review of the various normalization techniques can be found in the review by Quackenbush (2002)

Sample preparation

As already described, the nucleotidic material to be hybridized in the chip undergoes amplification/labeling step, obtained by some combination of reverse-transcriptase/PCR/*in vitro* transcription. This step can also represent a source of noise, as for example not all mRNA are reverse transcribed with the same efficiency. As this effect is gene-specific, the fluorescence intensity that is measured for a gene at the end of a study may not be a true reflection of its original mRNA level. This problem prevents the data analyst from comparing fluorescence intensities for different genes across a single sample. Nevertheless, even in the presence of this problem, intensity data on one gene can still be compared across samples.

Experimental design

Experimental design is crucial for the effects it has on the final data interpretation. Different designs have been proposed (Churchill, 2002) and the final choice depends on the particular experimental conditions and the resources availability.

Depending on the scientific problem addressed and on the limits of the technology used, more emphasis can be put on the control of the technical biases, which implies more technical replicates (e.g., replicates of the same sample on more chips) or to the control of the biological variation across the samples, implying the assay of many biological replicates (e.g., tumor samples extracted from different individuals).

Large number of genes

Microarrays are normally used to monitor the expression of thousands of genes at the same time. This feature represents obviously a great opportunity but poses also serious challenges to data analysis. In many cases what could look exciting at a first glance, like for example a group of genes of interest being differentially expressed in two biological conditions, can just be the result of a random phenomena due to the high dimensionality of the data. Rigorous statistical techniques need to be applied indeed to guarantee the soundness of the observations and the possibility to generalize them.

Biological factors

Expression profile studies are in many cases based on the implicit assumption that the concentration of each protein in the cell, which is the functional product of a gene, is determined by the abundance of the corresponding mRNA. Nevertheless it is important to remember that the composition of proteins in a cell can only partially ascribed to the levels of the corresponding mRNAs. This means that if a gene is highly transcribed, the corresponding protein is not necessarily present at high concentration because of the several mechanisms controlling gene expression. Moreover, even assuming that the amount of the protein is directly proportional to the amount of mRNA, the proteins do often require a number of post-translational modifications in order to become active and exert their role in the cell.

Array quality assessment

A fundamental aspect of data analysis is the evaluation of the microarray experiment quality. Sources of technical variation could be identified and be related to a particular chip region (topological variations), the composition of the probes (GC content), dye inefficiency, etc.

Defining metrics for quality assessment allows discarding the data coming from below standard arrays in the very initial phase of the analysis and avoids misleading results.

Microarray data analysis

Given the high volume of information generated by microarrays and in the light of the issues described in the previous section, the usage of computational analytical methods is essential for the management and the interpretation of the data (Allison *et al.*, 2006) (Fig. 1.3). The analytical components of a microarray expression profiling experiment can be divided into:

- Design: set-up of the optimal experimental plan to minimize the "noise" and maximize the quality of the information derived (Churchill, 2002).
- Preprocessing: processing of the microarray image and normalization of the data to correct for technical sources of variability (Schuchhardt, 2000). Preprocessing steps can also include data transformation, data filtering (Pounds *et al.*, 2005) and background subtraction.
- Inference: testing of statistical hypothesis normally used to identify the genes differentially expressed in groups of samples.
- Clustering: genes/samples division into classes (D'Haeseleer, 2005), pathway analysis (Chapter 3).
- Validation of findings: application of statistical techniques to confirm the observations made in the study.

This section gives an overview of clustering, which is the analytical component more concerned with the extraction of biological information out of the expression data.

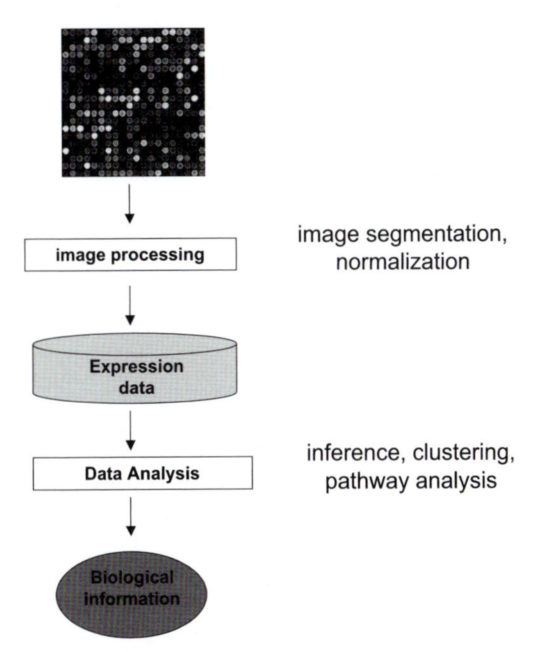

Figure 1.3 Scheme of a typical data analysis workflow. The images produced upon scanning are processed and data are normalized to correct for technical sources of variability. The resulting gene expression data can then be stored in a database and analyzed by pathway analysis, clustering and inference techniques to extract relevant biological information. This figure is also reproduced in color in the color section at the end of the book.

Clustering methods aim at classifying genes and samples, allowing a more powerful and structured data exploration than what would be achieved by an individual gene-by-gene or sample-by-sample analysis. The goal of clustering is indeed to subdivide a set of genes (or samples) in such a way that similar items fall into the same cluster, whereas dissimilar items fall in different clusters.

Current clustering methodologies can be divided into two categories: "unsupervised" approaches, which do not require the *a priori* input or knowledge of the information contained in the data, and "supervised" approaches, which determine genes and/or samples that fit a predetermined pattern, based on *a priori* knowledge.

Unsupervised clustering methods

As we have seen, the true power of microarray analysis comes from the analysis of many genes in many different samples to identify patterns of gene expression across different biological conditions. The largely adopted "guilt by association" paradigm described above is based on the assumption that genes participating in a common pathway, or involved in the response to a common environmental challenge, should be regulated by coordinated expression levels. In this light, one of the most important goals of data analysis is that of identifying genes that show similar patterns of expression across samples. Similarly, the analysis can be directed towards the identification of biological samples sharing similar expression of determined gene categories. To address this task, there exists a large group of statistical methods, generally referred to as "unsupervised clustering methods."

The two most common classes of unsupervised clustering methods are hierarchical clustering and partitioning.

Hierarchical clustering

In hierarchical clustering, genes (or samples) are grouped in a hierarchical way: each cluster is subdivided into smaller clusters, forming a tree-shaped data structure or dendrogram.

The method starts considering each object (gene or sample) as a cluster. In each of the subsequent steps, the two closest clusters are merged into one cluster and the distances between the clusters are recalculated; the process proceeds until there is only one cluster left. Intercluster distance can be defined in different ways; some of the most common are "single linkage" (the shortest distance from any member of one cluster to any member of the other cluster), "complete linkage" (longest distance from any member of one cluster to any member of the other cluster), "average linkage"/UPGMA (unweighted pair-group method: average of the distances from any member of one cluster to any member of the other cluster), and "centroid linkage"/UPGMC (unweighted pair-group method: centroids; distance between the cluster centroids).

The result of the clustering is usually represented by a dendrogram where the length of the arms connecting to nodes (or cluster of points) represents the distance between the genes or samples represented by the nodes (Fig. 1.4). The same procedure can be independently applied to cluster the genes or the samples. When both the genes and the samples are clustered using a hierarchical clustering algorithm, the result can be represented graphically by an ordered *heatmap* diagram (Fig. 1.4).

In this commonly used way to represent expression data rows and columns are reordered according to the tree defined by the hierarchical clustering algorithm. As a result, the proximity of two genes in one dimension (vertical dimension in Fig. 1.4) indicates similarity in their expression pattern across the various samples assayed. In the same way proximity of the samples represented on the orthogonal dimension (horizontal dimension in Fig. 1.4) indicates general transcriptional similarities between the samples. The colors indicate the expression values of the genes: in green are usually the genes down regulated and in red the genes upregulated.

One limitation of this approach is that the expression pattern similarities can often result from a general non-specific biological effect such as the cellular stress; this can cause different biological samples to appear similar and subtle gene expression differences be obscured by global, not functionally specific signatures.

Partitioning methods

Partitioning methods, such as "self-organizing maps" (SOMs) or "*k*-means" subdivide the data into a number of subsets predetermined by the analyst, without any implied hierarchical relationship between these clusters. For the *k*-means method, the algorithm starts with *k* randomly chosen cluster centroids, and each object is assigned to the cluster with the closest centroid. Next, the centroids are reset to the average of the genes in each cluster. This process proceeds until no more genes change cluster. Different initial centroid positions may yield different cluster results, and it is important to run the algorithm several times with different random seeds. Moreover, as the definition of the right number of clusters in the data is often an issue, a common approach is to try the clustering with different numbers of clusters and select a posteriori the optimal number of clusters. Fig. 1.5 shows an example of the gene clusters created applying the algorithm with $k = 4$ on an arbitrary expression dataset.

Similarity distances

In order to evaluate the similarity of genes, whose expression has been measured on several samples, a distance measure among genes needs to be defined.

Each gene expression profile can be represented as a point in the *N*-dimensional space where the number of dimensions is equal to the number of samples.

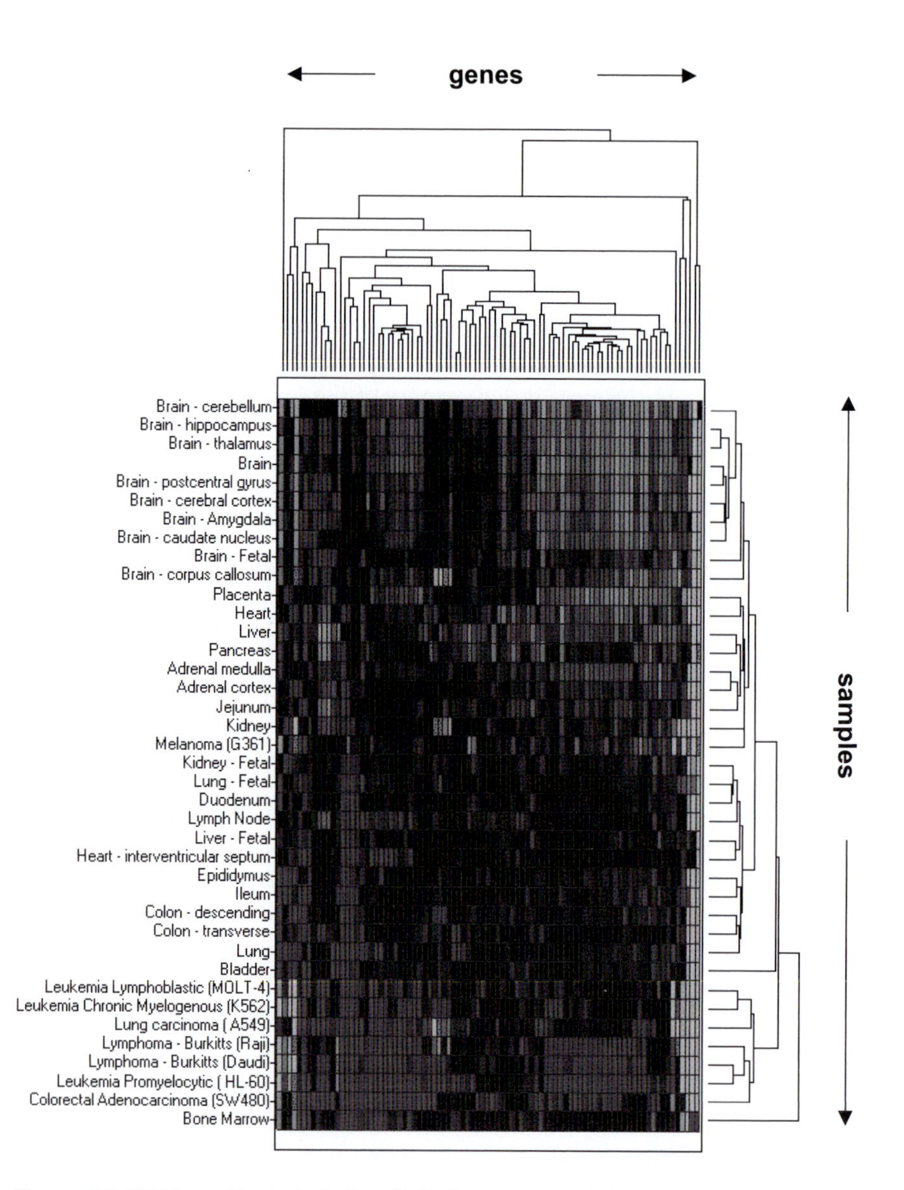

Figure 1.4 2D hierarchical clustering. Both the genes and the samples are clustered using a hierarchical clustering algorithm. The result can be represented graphically by an ordered *heatmap* diagram. Data rows (genes) and columns (samples) are reordered according to the tree defined by the hierarchical clustering algorithm. As a result, the proximity of two genes in one dimension (vertical dimension) indicates similarity in their expression pattern across the various samples assayed and the proximity of the samples represented on the orthogonal dimension (horizontal dimension) indicates general transcriptional similarities between the samples. The colors indicate the expression values of the genes: in green are usually the genes down regulated and in red the genes upregulated with respect to a baseline sample. This figure is also reproduced in color in the color section at the end of the book.

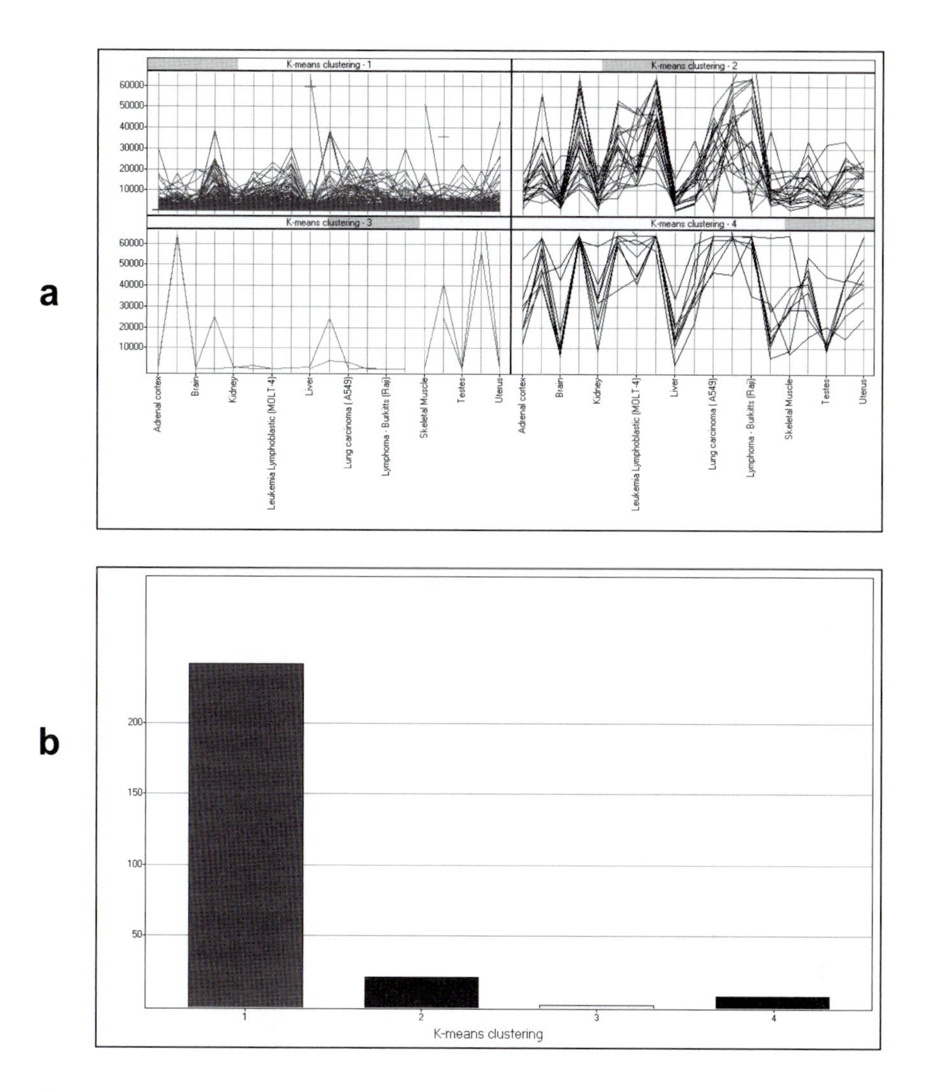

Figure 1.5 *k*-means clustering. Example of the gene clusters created applying the algorithm with $k = 4$ on an arbitrary expression dataset. (a) Four distinct gene expression cluster created by the *k*-means algorithm. Each panel shows the expression profile of the cluster genes across the samples used for profiling. (b) The number of genes falling into each different cluster can be represented by an histogram. This figure is also reproduced in color in the color section at the end of the book.

The similarity between two genes can than be expressed as the Euclidean distance between the two points. The same method can be applied to calculate the similarity of the samples, according to the expression of the genes in each sample. In this case a sample is represented by a point in the N-dimensional space where N is the number of genes.

The Euclidean distance between two points a and b in the N-dimensional space is defined as:

$$\sqrt{\sum_{i=1}^{N}(a_i - b_i)^2}$$

where N is the number of dimensions (Fig. 1.6a).

Another widely used way to measure the profile similarity is the Pearson's correlation distance. As correlation distance is not sensitive to scaling and differences in average expression level (whereas Euclidean distance is), this measure is appropriate when one wants to compare the shape of two profiles (the expression trends among the samples of different genes), rather than the gene expression absolute values (Fig. 1.6b).

The correlation between two profiles, a and b, with N dimensions is calculated as

$$\frac{\text{cov}(a,b)}{std(a) \cdot std(b)}$$

where COV is the covariance of the two profiles and STD is the standard deviation of each profile.

Supervised clustering methods

Supervised clustering methods (often also referred as "machine learning" methods) assign objects to *a priori*-defined categories. Algorithms are typically developed and evaluated on a "training" dataset and an independent "test" dataset,

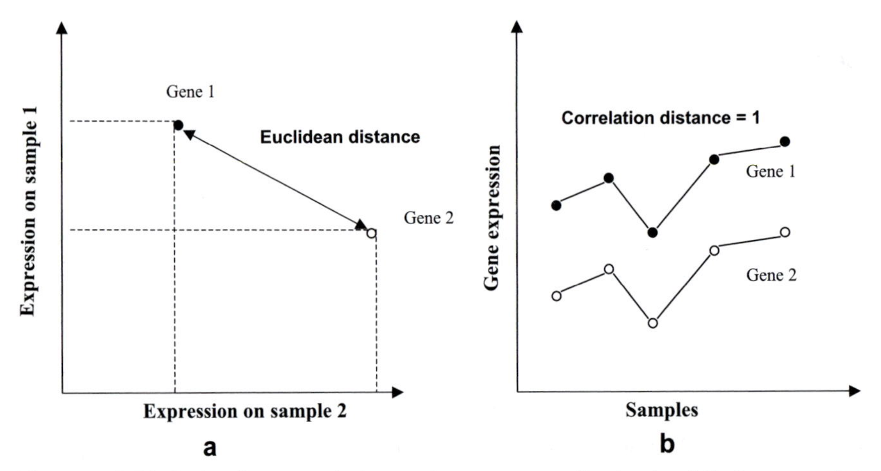

Figure 1.6 Similarity distances between two genes on the basis of their expression profiles. (a) Euclidean distance between two genes, measured on two samples; each axis represents the gene expression value on a sample. The distance between the two genes is given by the geometrical distance of the two points representing the genes in the N-dimensional space, where N is the number of samples. (b) Profiles of two genes measured on five samples have identical shape and therefore maximum correlation (equal to 1).

respectively, in which the categories to which objects belong are known before they are used in practical applications. For example, by measuring gene-expression patterns using RNAs collected from various patients for which disease-stage classification or survival data are available, microarray data can be used to "train" an algorithm that can then be applied to the classification of other previously unclassified samples. This allows the definition of a "molecular expression fingerprint" for disease-stage diagnosis or for prognosis. Similarly supervised methods could be applied to verify the existence of a gene expression fingerprint associated with the response to a drug. The molecular profile could then be used to distinguish good from bad responders to a given drug treatment.

Widely used supervised clustering methods include "support vector machine" (SVM), "random forests," and "neural networks."

An extensive illustration of the application of supervised methods for toxicogenomics studies and for the classification of biological samples can be found respectively in Chapters 2 and 6 of this book.

Public resources for expression data

Publicly available microarray data, analysis tools and useful information can be found at the following links:

Stanford Microarray Database (SMD) (http://www.dnachip.org/)

SMD stores raw and normalized data from microarray experiments, as well as their corresponding image files. In addition, SMD provides interfaces for data retrieval, analysis and visualization.

Gene Expression Omnibus (GEO) (http://www.ncbi.nlm.nih.gov/geo/)

The GEO serves as a public repository for a wide range of high-throughput experimental data. These data include single- and dual-channel microarray-based experiments measuring mRNA, genomic DNA and protein abundance, as well as non-array techniques such as serial analysis of gene expression (SAGE), and mass spectrometry proteomic data.

Microarray Gene Expression Data (MGED) (http://www.mged.org/)

The MGED Society is an international organization of biologists, computer scientists, and data analysts that aims to facilitate the sharing of microarray data generated by functional genomics and proteomics experiments.

ArrayExpress (http://www.ebi.ac.uk/arrayexpress/)

ArrayExpress is a public repository for microarray data. Besides the browsing and querying of the data it offers the possibility to export the data for further analysis with the EBI's free analysis software.

Expression Profiler
(http://www.ebi.ac.uk/microarray/ExpressionProfiler/ep.html)
Expression Profiler is a set of tools for clustering, analysis and visualization of gene expression and other genomic data.

Book's content

Almost all the chapters of this book refer to gene expression studies; in fact these studies still dominate the microarray usage area, although an increasing number of alternative applications is emerging. Being a field in continuous evolution and because of the multiplicity of variations on the "microarray theme," we do not intend to cover here the complete spectrum of DNA microarrays potentialities. The arguments we propose are selected among the research areas in which microarray approaches have not only shown their "promises" but have also concretely demonstrated their usefulness and success.

It will be illustrated how microarrays can assist basic research and drug discovery by the generation of "body maps," maps of gene expression in different normal tissues (chapter 5), the reconstruction of gene networks (Chapter 2) and the study of biochemical pathways (Chapter 3). Some of the concepts mentioned here about the clustering of biological samples will be explained in more details in a chapter dedicated to supervised classification methods (Chapter 6) and a real case study on the transcriptional effects of vaccine treatment will be extensively described (Chapter 7). A final chapter will be dedicated to the illustration of a large spectrum of "not-gene-expression" applications of DNA microarray (Chapter 8).

Going through these themes we hope to show the great extent to which different aspects of microarray studies are connected to each other and how multiple approaches can be used towards the reconstruction of a complex "molecular picture" of the cell.

References

Ahmadi, K.R., Weale, M.E., Xue, Z.Y., Soranzo, N., Yarnall, D.P., Briley, J.D., Maruyama, Y., Kobayashi, M., Wood, N.W., Spurr, N.K., Burns, D.K., Roses, A.D., Saunders, A.M., and Goldstein D.B. (2005). A single-nucleotide polymorphism tagging set for human drug metabolism and transport. A single-nucleotide polymorphism tagging set for human drug metabolism and transport. Nat. Genet. 37, 84–89.

Ahrendt, S.A., Halachmi, S., Chow, J.T., Wu, L., Halachmi, N., Yang, S.C., Wehage, S., Jen, J., and Sidransky, D. (1999). Rapid p53 sequence analysis in primary lung cancer using an oligonucleotide probe array. Proc. Natl. Acad. Sci. USA 96, 7382–7387.

Alizadeh, A.A., Eisen, M.B., Davis, R.E., Ma, C., Lossos, I.S., Rosenwald, A., Boldrick, J.C., Sabet, H., Tran, T., Yu, X., Powell, J.I., Yang, L., Marti, G.E., Moore, T., Hudson, J. Jr., Lu, L., Lewis, D.B., Tibshirani, R., Sherlock, G., Chan, W.C., Greiner, T.C., Weisenburger, D.D., Armitage, J.O., Warnke, R., Levy, R., Wilson, W., Grever, M.R., Byrd, J.C., Botstein, D., Brown, P.O., and Staudt, L.M. (2000). Distinct types of diffuse large B-cell lymphoma identified by gene expression profiling. Nature *403*, 503–511.

Allison, D.B., Cui, X., Page, G.P., and Sabripour, M. (2006). Microarray data analysis: from disarray to consolidation and consensus. Nat. Rev. Genet. *7*, 55–65.

Behr, M.A., Wilson, M.A., Gill, W.P., Salamon, H., Schoolnik, G.K., Rane, S., and Small, P.M. (1999). Comparative genomics of BCG vaccines by whole-genome DNA microarray. Science *284*, 1520–1523.

Buck, M.J., and Lieb, J.D. (2004). ChIP-chip: considerations for the design, analysis, and application of genome-wide chromatin immunoprecipitation experiments. Genomics *83*, 349– 360.

Churchill, G.A. (2002). Fundamentals of experimental design for cDNA microarrays. Nat. Genet. *32*, Suppl., 490–495.

Cronin, M.T., Fucini, R.V., Kim, S.M., Masino, R.S., Wespi, R.M., and Miyada, C.G. (1996). Cystic fibrosis mutation detection by hybridization to light-generated DNA probe arrays. Hum. Mutation *7*, 244–255.

Cutler, D.J., Zwick, M.E., Carrasquillo, M.M., Yohn, C.T., Tobin, K.P., Kashuk, C., Mathews, D.J., Shah, N.A., Eichler, E.E., Warrington, J.A., and Chakravarti, A. (2001). High-throughput variation detection and genotyping using microarrays. Genome Res. *11*, 1913–1925.

D'Haeseleer, P. (2005). How does gene expression clustering work? Nat. Biotechnol. *23*, 1499–1501.

di Bernardo, D., Thompson, M.J., Gardner, T.S., Chobot, S.E., Eastwood, E.L., Wojtovich, A.P., Elliott, S.J., Schaus, S.E., and Collins, J.J. (2005). Chemogenomic profiling on a genome-wide scale using reverse-engineered gene networks. Nat. Biotechnol. *23*, 377–383.

Draghici, S. (2003). Data Analysis Tools for DNA Microarrays. Chapman & Hall/ CRC.

Hacia, J.G. (1999). Resequencing and mutational analysis using oligonucleotide micro-arrays. Nat. Genet. *21*, 42–47.

Hughes, T.R., Marton, M.J., Jones, A.R., Roberts, C.J., Stoughton, R., Armour, C.D., Bennett, M.A., Coffey, E., Dai, H., He, Y.D., Kidd, M.J., King, A.M., Meyer, M.R., Slade, D., Lum, P.Y., Stephaniants, S.B., Shoemaker, D.D., Gachotte, D., Chakraburtty, K., Simon, J., Bard, M., and Friend, S.H. (2000). Functional discovery via a compendium of expression profiles. Cell *102*, 109–126.

Hughes, TR, Mao M, Jones AR, Burchard J, Marton MJ, Shannon, K.W., Lefkowitz, S.M., Zinan, M., Schelter, J.M., Meyer, M.R., Kobayashi, S., Davis, C., Dai, H., He, Y.D., Stephaniants, S.B., Cavet, G., Walker, W.,L., West, A., Coffey, E., Shoemaker, D.D., Stoughton, R., Blanchard, A.R., Friend, S.H., and Linsley, P.S. (2001). Expression profiling using microarrays fabricated by an ink-jet oligonucleotide synthesizer. Nat. Biotechnol. *19*, 342–347.

Iyer, V., *et al.* (2001). Genomic binding sites of the yeasT-cell-cycle transcription factors SBF and MBF. Nature *409*, 533–538.

Kary, L. (1997). The Mathematical Intelligencer, 19, 9–22

Kim, S.K., Lund, J., Kiraly, M., Duke, K., Jiang, M., Stuart, J.M., Eizinger, A., Wylie, B.N., and Davidson, G.S. (2001). A gene expression map for Caenorhabditis elegans. Science *293*, 2087–2092.

Lucito, R., Healy J., Alexander J., Reiner A., Esposito D., Chi, M., Rodgers, L., Brady, A., Sebat, J., Troge, J., West, J.A., Rostan, S., Nguyen, K.C., Powers, S., Ye, K.Q., Olshen, A., Venkatraman, E., Norton, L., and Wigler, M. (2003). Representational

oligonucleotide microarray analysis: a high-resolution method to detect genome copy number variation. Genome Res. 13, 2291–2305.
Martone, R., *et al.* (2003). Distribution of NF-kappaB-binding sites across human chromosome 22. Proc. Natl. Acad. Sci. USA *100*, 12247–12252.
Mischel, P.S., Cloughesy, T.F., and Nelson, S.F. (2004). DNA-microarray analysis of brain cancer: molecular classification for therapy. Nat. Rev. Neurosci. 5, 782–792.
Mootha, V.K., Lindgren, C.M., Eriksson, K.F., Subramanian, A., Sihag, S., Lehar, J., Puigserver, P., Carlsson, E., Ridderstrale, M., Laurila, E., Houstis, N., Daly, M.J., Patterson, N., Mesirov, J.P., Golub, T.R., Tamayo, P., Spiegelman, B., Lander, E.S., Hirschhorn, J.N., Altshuler, D., and Groop, L.C. (2003). PGC-1alpha-responsive genes involved in oxidative phosphorylation are coordinately downregulated in human diabetes. Nat. Genet. *34*, 267–273.
Pounds, S., and Cheng, C. (2005). Statistical development and evaluation of microarray gene expression data filters. J. Comput. Biol. *12*, 482–495.
Quackenbush, J. (2002). Microarray data normalization and transformation. Nat. Genet. *32*, Suppl., 496–501.
Quackenbush, J. (2003). Genomics. Microarrays—guilt by association. Science *302*, 240–241.
Raghuraman, M.K., Winzeler, E.A., Collingwood, D., Hunt, S., Wodicka, L., Conway, A., Lockhart, P.J., Davis, R.W., Brewer, B.J., and Fangman, W.L. (2001). Replication dynamics of the yeast genome. Science *294*, 115–121.
Ramaswamy, S., Tamayo, P., Rifkin, R., Mukherjee, S., Yeang, C.H., Angelo, M., Ladd, C., Reich, M., Latulippe, E., Mesirov, J.P., Poggio, T., Gerald, W., Loda, M., Lander, E.S., and Golub, T.R. (2001). Multiclass cancer diagnosis using tumor gene expression signatures. Proc. Natl. Acad. Sci. USA *98*, 15149–15154.
Rodriguez, B.A., and Huang T.H. (2005). Biochem Tilling the chromatin landscape: emerging methods for the discovery and profiling of protein–DNA interactions. Cell Biol. *83*, 525–534.
Schena, M. (2000). Microarray Biochip Technology. Eaton Publishing, Sunnyvale, CA.
Schuchhardt, J., Beule D., Wolski E. And Eickhoff H. (2000). Normalization strategies for cDNA microarrays. Nucleic Acids Res. *28*, e47i–e47v.
Shoemaker, D.D., Schadt, E.E., Armour, C.D., He, Y.D., Garrett-Engele, P., *et al.* (2001). Experimental annotation of the human genome using microarray technology. Nature *409*, 922–927.
Stoughton, R.B. (2005). Application of DNA Microarrays in Biology. Annu Rev Biochem.;74, 53–82
Smoot, J.C., Barbian, K.D., Van Gompel, J.J., Smoot, L.M., Chaussee, M.S., *et al.* (2002). Genome sequence and comparative microarray analysis of serotype M18 group A *Streptococcus* strains associated with acute rheumatic fever outbreaks. Proc. Natl. Acad. Sci. USA 99, 4668–4673.
van't Veer, L.J., Dai, H., van de Vijver, M.J., He, Y.D., Hart, A.A., *et al.* (2002). Gene expression profiling predicts clinical outcome of breast cancer. Nature 415, 530–536.
Wang, D.G., Fan, J.B., Siao, C.J., Berno, A., Young, P., Sapolsky, R., Ghandour, G., Perkins, N., Winchester, E., Spencer, J., Kruglyak, L., Stein, L., Hsie, L., Topaloglou, T., Hubbell, E., Robinson, E., Mittmann, M., Morris, M.S., Shen, N., Kilburn, D., Rioux, J., Nusbaum, C., Rozen, S., Hudson, T.J., Lipshutz, R., Chee, M.,and Lander, E.S. (1998). Large-scale identification, mapping, and genotyping of single-nucleotide polymorphisms in the human genome. Science *280*, 1077–1082.
White, K.P., Rifkin, S.A., Hurban, P., Hogness, D.S. (1999). Microarray analysis of *Drosophila* development during metamorphosis. Science *286*, 2179–2184.

Wikman, F.P., Lu, M.L., Thykjaer, T., Olesen, S.H., Andersen, L.D., Cordon-Cardo, C., and Orntoft, T.F. (2000). Evaluation of the performance of a p53 sequencing microarray chip using 140 previously sequenced bladder tumor samples. Clin. Chem. *46*, 1555–1561.

Gene Networks and Application to Drug Discovery 2

A. Ambesi, M. Bansal, G. Della Gatta, and Diego di Bernardo

Abstract

The drug discovery process is complex, time-consuming, and expensive, and includes preclinical and clinical phases. The pharmaceutical industry is moving from a symptomatic relief focus toward a more pathology-based approach, in which a better understanding of the pathophysiology should help deliver drugs whose targets are involved in the causative processes underlying the disease. Computational biology and microarray technology have the potential not only to speed up the drug discovery process, thus reducing the costs, but also to change the way drugs are designed. In this chapter we focus on the different computational and bioinformatics approaches that have been proposed and applied to the different steps involved in the drug development process, with particular emphasis on gene network models. The development of "gene-network reconstruction" methods is now making it possible to infer a detailed map of the regulatory circuit among genes, proteins and metabolites. It is likely that the development of these technologies will radically change, in the next decades, the drug discovery process as we know it today.

Introduction

The drug discovery process is complex, time-consuming, and very expensive. Typically, the time to develop a candidate drug is about 5 years, while the clinical phases leading, possibly, to the commercial availability of the drug are even longer (> 7 years) for a total cost of more than 700 million dollars (DiMasi *et al.*, 2003). The drug discovery process begins from the identification of an area of "unmet medical needs" and then proceeds by identifying "druggable" biological targets that could relief the symptoms of the disease, or, as in the recent years, that are involved in the causative process of the disease. The pharmaceutical industry is moving from a symptomatic relief focus towards a more pathology-based approach where a better understanding of the pathophysiology should help deliver drugs whose targets are directly involved in the causative processes underlying the disease (Ratti and Trist, 2001). The drug discovery process is very similar across different pharmaceutical companies. It consists of preclinical and clinical phases. In the *target identification and validation* step, "druggable" biological targets are identified. In the *hit identification* step, library of compounds ranging from tens to hundreds of thousands of compounds are screened against the "druggable" targets to identify those compounds that "hit" the targets using high-throughput screening (HTS). HTS methods based on experimental assays are reviewed extensively elsewhere (Hart, 2005). The number of compounds selected after this step is in the order of hundreds. By analyzing the structure of the selected compounds and identifying common active substructures, novel compounds containing those substructures are synthesized to significantly lower the number of lead compounds. This step is called *lead identification*. Structural bioinformatics and chemical informatics approaches to drug discovery are particularly useful in this step, however, widely used methods like structure–activity relationship (SARs) are outside the scope of this chapter. We refer the interested reader to Bredel and Jacoby (2004) and Fagan and Swindells (2000).

The leads identified are further refined to comply with pharmacokinetic constraints such as absorption and bioavailability, and to increase their potency and efficacy, while decreasing side effects and toxicity. This step is called *lead optimization*. Knowledge of the mode of action (MOA), that is, the identification of the therapeutic molecular target of the drug, can simplify the task of optimizing the drug candidate. Understanding the MOA can help predicting the effect of drug interactions and allow SARs to guide medicinal chemistry efforts toward optimization (Hart, 2005). However, for many drugs, the targets are unknown and difficult to find among the thousands of gene products in a typical genome.

Many new compounds fail when they are tested in humans due to lack of efficacy. Testing for efficacy early during the drug discovery process (i.e., before the clinical phases) is essential for reducing costs and time required. Therefore, the development of experimental and computational approaches to test for efficacy *in vitro* is critical.

After the preclinical phases, a candidate compound is then selected and the clinical phase of the process can begin. This consists of clinical phase I, phase II, and phase III and possibly the launch into the market. Many compounds fail in the clinical phases of the process thus leading to consistent waste of time and money. A good review of the evolution of the drug discovery process can be found in Ratti and Trist (2001).

Computational biology and microarray technology have the potential not only of speeding up the drug discovery process thus reducing the costs, but also of changing the way drugs are designed. In this chapter we will focus on the different approaches that have been proposed and applied to the different steps involved in drug development as shown in Table 2.1, with particular emphasis to the application of gene networks models inferred from microarray data. Our aim is to describe the different computational methods that have been used so far to tackle these problems by giving examples of applications. Since we cannot

Table 2.1 Classification of the reviewed manuscripts according to the computational methods used and their application to the drug discovery process

Drug discovery		Classifiers	Network/pathway reconstruction
Target identification and validation		Stoughton and Friend (2005) (review), Walker (2001) (review), Hughes *et al.* (2000), Gasch *et al.* (2000), Stegmeir *et al.* (2004), Brown *et al.* (2000)	Gardner *et al.* (2003), Basso *et al.* (2005), Gardner *et al.* (2005) (review) (Gardner and Faith (2005), Apic *et al.* (2005) (review)
Hit identification, lead identification and optimization	Mode of action (MOA)	Perlman *et al.* (2004), Parsons *et al.* (2003, 2004), Marton *et al.* (1998), Giaever *et al.* (1999, 2004), Lum *et al.* (2004); Hughes *et al.* (2000), Betts *et al.* (2003), Paull *et al.* (1989, Weinstein *et al.* (1997), Bao *et al.* (2002)	di Bernardo *et al.* (2005), Imoto *et al.* (2003), Haggarty *et al.* (2003)
	Efficacy and toxicity	Bugrim *et al.* (2004) (review), Bugrim *et al.* (2004), Szakacs *et al.* (2004), Staunton *et al.* (2001), Scherf *et al.* (2000), Gunther (*et al.* (2003), Gunther *et al.* (2005), Hamadeh (2002a,b), Dan *et al.* (2002)	Not known

be comprehensive, we tried to compensate for this by referring the interested readers to published reviews that have been written on this subject. The organization of this chapter is based on classifying drug discovery approaches into two major categories. The first section of this chapter reviews classifier-based algorithms, which try to determine drug specific patterns as biomarkers of a compound activity, while the second section assesses more complex methods that attempt to infer the network of gene–gene interactions that are perturbed by a drug. We further subdivided those sections in subsections, each focusing on specific steps of the drug discovery process.

Classifier-based algorithms

A classifier is an algorithm that uses a set of input or predictor variables $x = (x_1, x_2, \ldots, x_n)$ to predict one or more response variables $y = (y_1, y_2, \ldots, y_m)$ (Fig. 2.1). For example, x can be a set of measurements of the expression of n genes in response to a drug treatment in a tumor cell type and y can represent the efficacy of the drug for that tumor cell type. Classifiers can be further subdivided in supervised learning methods and unsupervised learning methods. In supervised learning a training set of "solved cases" is used to train a model to recognize what will be the response y given the input variables x. Supervised learning methods may be thought of as "learning with a teacher models" in

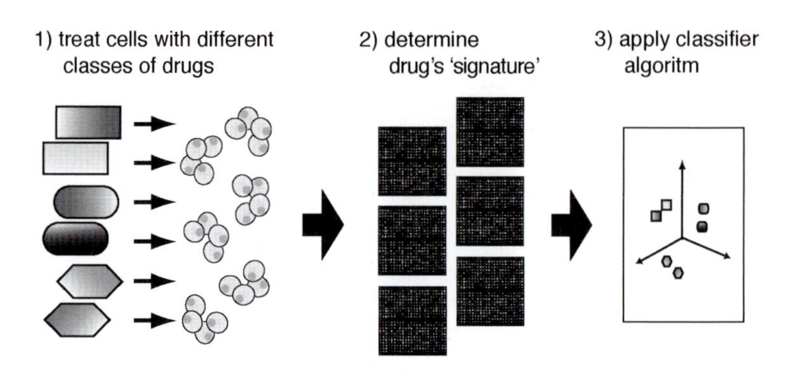

Figure 2.1 Classifier-based algorithms.

which a student gives an answer $\hat{y}$ to each question x in the training set, and the teacher provides the correct answer y. After the training, the student should be able to give the correct answer to a new question that was not in the training set. If y and $\hat{y}$ are coded as numerical values, we can define a loss function $L(y,\hat{y})$, for example, $L(y,y^{\wedge}) = (y - \hat{y}(\theta))^2$ where θ are the parameters of the model to be learned. By minimizing this function over θ on the training set, one finds the values of the model parameters θ. For example, linear discriminant analysis (LDA) is a supervised learning in which $\hat{y} = \theta x$.

In unsupervised learning, or "learning without a teacher," one has a set of n observations $x = (x_1, x_2, \ldots, x_n)$ without the correct response variables. Cluster analysis is an example of unsupervised learning method whose goal is to group a collection of objects into subsets or "clusters," such that the objects within each cluster are more closely related to one another than those assigned to different clusters. In addition the goal can also be to arrange the clusters in a natural hierarchy. A commonly used hierarchical clustering is the one described by Eisen *et al.* (1998). Unsupervised methods have the advantage that they are "data driven" and do not rely on *a priori* knowledge. A comprehensive and detailed description of these methods can be found in the excellent book by Hastie *et al.* (2001).

Target identification and validation

Whole-genome gene expression data, proteomic data or metabolomic data, also named "molecular profiling" in a recent review (Stoughton and Friend, 2005), can be used to build classifier algorithms able to help in the process of identifying "druggable" gene/protein/metabolite targets.

An example of an unsupervised learning method can be found in Hughes *et al.* (2000). These authors constructed a reference database of whole-genome expression profiles referred to as a gene expression "compendium" generated by 300 diverse mutations and chemical treatments in *Saccharomyces cerevisiae*. A 2D hierarchical clustering (Eisen *et al.*, 1998; Hartigan, 1975) was used to cluster genes and experiments using as the similarity measure the correlation coefficient. Genes and experiments were reordered according to the resulting clustering similarity trees. By examining the clusters the authors were able to find an unknown ORFs that clustered among genes involved in the ergosterol biosynthesis and experiments that were perturbing this pathway, thus deducing these ORFs to belong to this pathway. They then experimentally confirmed that eight of these ORFs were indeed required for sterol metabolism. Since sterol metabolism is a "druggable" pathway in yeast for antimycotic drugs, this work shows how novel targets can be identified via bioinformatics approaches. A similar method has been applied by Gasch *et al.* (2000), who performed a hierarchical clustering of 142 whole-genome arrays in *S. cerevisiae* in response to environmental changes and were able to clarify the regulation mechanisms in which three transcriptions factors were involved.

An example of supervised learning for understanding the function of gene from gene expression data is given in Brown *et al.* (2000), in which support vector machines (SVMs) (Hastie *et al.*, 2001) are used. When applied to gene expression data, an SVM begins with a set of genes that have a common function, for example genes coding for ribosomal proteins or genes coding for components of the proteasome. In addition, a separate set of genes that are known not to be members of the functional class is specified. These two sets of genes are combined to form a set of training examples in which the genes are labeled positively if they are in the functional class and are labeled negatively if they are known not to be in the functional class. By analyzing expression data from 2,467 genes from the budding yeast *S. cerevisiae* measured in 79 different DNA microarray hybridization experiments the authors were able to correctly assign genes to five functional classes from the Munich Information Center for Protein Sequences Yeast Genome Database (MYGD). The method is compared with hierarchical clustering and shown to be marginally better, but this could have been expected since supervised learning methods have access to additional information as provided by the training set.

An original high-throughput drug screening strategy based on unsupervised learning is used by Segmaier *et al.* (2004), which, unlike most commonly used methods, does not simply screen for compounds that interact with specific molecular targets. The authors preliminarily define a gene expression signature for the target post-treatment phenotype, or "cellular state" of interest. Specifically, in this study the target cellular state was differentiated neutrophils and monocytes from control individuals versus pretreatment bone marrow samples derived from acute myelogenous leukemia (AML) patients. A "handful" of marker genes were selected, unfortunately not in a generalized manner but rather arbitrarily, from the differentiation-correlated genes. Those markers were then used to develop a detection assay called gene expression-based high-throughput screening (GE-HTS) based on multiplexed RT-PCR and single base extension (SBE) reaction followed by MALDI-TOF mass spectrometry. Eight target compounds identified by GE-HTS in this study were validated in several ways including morphological observations and functional measures. Interestingly, the broader cellular genetic program of differentiation beyond the selected handful of marker genes was also investigated, again through a correlation-based statistical test. The authors analyzed triplicate microarray expression data from HL-60 cell lines treated with eight different compounds. Six out of the eight expression profiles were found to be significantly similar to the gene expression differences characterizing the original AML-vs.-controls primary cells, as determined by the Mantel test (Stegmaier *et al.*, 2004). This test is an unbiased, global measure of similarity, and indicates that the six compounds induced a nonrandom pattern of gene expression consistent with differentiation. The advantage of GE-HTS is that the development of the assay does not require any specialized assays such as traditional methods based on antibodies or reporter constructs or cellular phenotypes, and, once the gene expression signature pattern is defined, the procedure is rather straightforward.

Although they may result in outstanding accuracy performance, correlation-based methods do not easily provide insight into the mechanisms of action common to the therapeutic category, but rather capture silent features of drug efficacy by their correlation to biomarker signatures based on gene expression patterns.

Hit identification, lead identification, and optimization: mode of action (MOA)

One of the first bioinformatics approaches to determine mode of action of a compound was based on a simple supervised learning approach (Paull *et al.*, 1989). In 1985 the National Cancer Institute (USA) established a primary screen in which compounds were tested *in vitro* for their ability to inhibit growth of 60 different human cancer cell lines (Weinstein *et al.*, 1997). To each compound tested it is possible to associate a value quantifying the differential growth inhibition (GI) for each cell line (treated vs. untreated). The algorithm developed by Allen and coworkers, named COMPARE, measures the similarity of the GI "signature" of a novel compound against a database of "signatures" of compounds with known MOA. The similarity is obtained simply by computing the average differences between the signatures of the test compound and each of the signatures in the database. Ranking according to this measure of similarity, one can infer the MOA of the novel compound as the one of the most similar compound in the database. An extension of this approach based on hierarchical clustering and integration of different dataset from the NCI 60 cell lines has been proposed by Weinstein *et al.* (1997). A more sophisticated approach using SVM to classify drugs into five mechanistic classes using drug activity profiles and the gene expression profiles of each of the untreated NCI 60 cell lines, has been proposed by Bao *et al.* (2002).

Unsupervised approaches have been applied extensively in this area. Marton *et al.* (1998) were pioneers of the "signature approach" based on gene expression profile following drug treatment. In this approach the drug signature is compared to a mutant strain signature using a correlation coefficient as a measure of similarity:

$$p = \frac{\sum x_k y_k}{\sqrt{\sum x_k^2 \sum y_k^2}}$$

They also proposed a further "decoder" step where the mutant strains whose expression profiles were most similar to the drug-treated cells are treated with the drug, generating an expression signature in the mutant strain. If the mutated gene encodes a protein involved in the pathway affected by the drug, then the signature in mutant cell should be different or, ideally, absent. Marton *et al.* (1998) did a proof of principle study on FK506 and the calcineurin signaling pathway as a model system.

The previously described work by Hughes *et al.* (2000), is another good example of how hierarchical clustering and correlation can be used for understanding the MOA of a drug. The authors used the gene expression "compendium" to identify the target of the commonly used topical anesthetic dyclonine. In order to find the target of the compound, the authors treated the yeast cells with the compound and compared the gene expression profile to the most similar expression profiles in the compendium using the correlation coefficient as the similarity measure. The erg2Δ strain (knock-out of the erg2 gene) was most similar to the dyclonine treatment thus suggesting, correctly as verified experimentally, that this gene is the molecular target of the drug. Since this gene is conserved in human but codes for the sigma receptor, a neurosteroid-interacting protein, the MOA of the drug in human has also been explained.

Hierarchical clustering methods have been applied not only to gene expression data, but also to chemical-genetic and genetic interaction data. Parsons *et al.* (2004) screened ~ 4,700 yeast deletion mutants for hypersensitivity to 12 diverse inhibitory compounds. Hypersensitivity was measured from digital images of plates by quantifying colony area growing in drug-medium versus no-drug control medium. Hypersensitive strains for a given drug were coded as 1, and with a 0 otherwise. These data (a vector of ~ 4,700 0s and 1s for each drug) were used for 2D hierarchical clustering. Both genes and compounds are clustered together upon the similarity of their chemical–genetic interactions. By analyzing the clusters they were able to detect genes whose deletion was associated with sensitivity to multiple compounds, thus enabling them to identify a multidrug-resistant gene set. To identify the mode of action of a compound, they performed synthetic lethal screens between ERG11 mutants and the ~ 4,700 deletion strains. The overlap between the genes that were synthetic lethal with ERG11 mutants, with the genes whose deletions were lethal after treatment with flucanozole, was used to infer the MOA of this drug.

Related to these methods are drug-induced haploinsufficiency screens first proposed by Giaever *et al.* (1999). Drug-induced haploinsufficiency occurs when lowering the dosage of a single gene from two copies to one copy in diploid cells results in a heterozygote that displays increased sensitivity to the drug as compared to the wild-type strain. These screens make use of a fitness defect score (Giaever *et al.*, 2004) that is computed using different methods (Baetz *et al.*, 2004; Lum *et al.*, 2004).

Hierarchical clustering has been applied also to data derived from automated microscopy in order to identify drug MOA. Perlman *et al.* (2004) chose 200 compounds, 90 of which were drugs with known MOA. They cultured HeLa (human cancer) cells in 384-well plates to near confluence, and treated them with 13 threefold dilutions of each drug for 20 hours, covering a final concentration range from micromolar to picomolar. They chose 11 distinct fluorescent probes covering a range of biological processes. Using automated fluorescence microscopy they measured for each cell, region and probe, a set of descriptors including size, shape, intensity, as well as ratios of intensities between regions for a total of 93 descriptors. For each descriptor they developed a titration-invariant similarity score (TISS) to allow comparison between dose-response profiles independent of starting dose. TISS scores for 61 compounds were computed and used for hierarchical clustering; the data matrix used for clustering consisted of 61 compounds by 93 TISS scores. Once again they found that drug with similar mechanism of action clustered together, thus allowing inference of drug MOA for drugs with unknown molecular targets.

Signature expression profiles were used by Betts *et al.* (2003) to determine the differential mode of action of three active drugs against *Mycobacterium tuberculosis*, and as a means of identifying novel and efficacy-optimized active drugs. In this study the authors show that although global response profiles of isoniazid and thiolactomycin are more closely related to each other than to that of triclosan, there are differences that distinguish the mode of action of these two drugs. A mathematical model is proposed to discriminate between the three compounds and also the vehicle control treatment. The main sources of variance of the data were obtained by principal component analysis (PCA). The principal components are a linear combination of all the gene intensities. Partial least squares discriminant analysis was performed on a subset of data selecting the dose and the time point that maximized separation of experimental groups. The 500 top-ranking genes thus identified were further processed by stepwise linear discriminant analysis in order to generate a mathematical model for the probability $P_i(x)$ of a gene expression signature x belonging to classification group i based on the following discriminant function:

$$P_i(x) = \frac{e^{D_i^2(x)}}{\sum_{j=1}^{n} e^{D_j^2(x)}} \qquad i = 1, 2, \ldots, n \tag{2.1}$$

where $D_i^2(x)$ is the discriminant score of the signature x for group i.

Methods that rely on a dataset for the construction of a classifier model, without implementing more robust statistical analyses, such as running a series of training and testing data in a "leave-one-out" manner, although accurately performing on the training dataset may lead to the construction of a model that "overfits" the data, and thus may not perform well on new data obtained using different treatments.

Hit identification, lead identification, and optimization: efficacy and toxicity

A large part of the efforts based on computational and bioinformatics approaches have been directed to predict sensitivity of cancer cell lines to different compounds. Sherf *et al.* (2000) aimed at relating sensitivity to therapy with gene expression using an unsupervised approach. They used the database of drug activity profiles (Growth Inhibition after 48 h of drug treatment) of more than 70,000 compounds on NCI 60 cell lines, together with gene expression profiles of 9,703 genes measured using cDNA microarrays for each of the 60 untreated cell lines. They then performed a hierarchical clustering of 118 compounds with known mechanism of action. In order to integrate drug activity profile with gene expression data, they chose Pearson correlation coefficient as a measure of similarity. This coefficient was calculated for each combination of a gene (expression profile across 60 cell lines) and a drug (GI activity profile across 60 cell lines). This yielded 1376 correlation coefficients for each of the 118 drugs. Using this technique they were able to associate sensitivity of leukemic lines to L-asparagine to the amount of asparagine synthetase. A similar technique has been proposed by Dan *et al.* (2002). They were able to identify gene markers for chemosensitivity for 55 anticancer drugs using gene expression data across 39 cell lines and drug activity profiles (GI). Similarly, Szakacs *et al.* (2004) and co-workers correlated expression profiles of all 48 human ABC transporters with patterns of drug activity in the NCI 60 cell lines. They were able to identify candidate substrates for several ABC transporters and compounds whose toxicities are potentiated by ABCB1-MDR1.

One potential application of microarrays in toxicology is their use in predicting toxicity of undefined chemicals by comparing their gene expression patterns in a biological model with databases of microarray-generated gene expression data corresponding to known toxicants. Feasibility of compound classification based on gene expression profiles is proven by several experiments. Hammadeh *et al.* (2002a,b), for example, analyzed rat liver gene expression patterns elicited by peroxisome proliferators, and enzyme inducers. These authors used several computational analyses including hierarchical clustering (Eisen *et al.*, 1998), PCA, pairwise Pearson correlation of gene expression profiles, and finally a combination of a genetic algorithm and K-nearest neighbor (GA/KNN) (Li *et al.*, 2001). Their results confirm that compound classification based on gene expression is feasible, and showed both strong within-class correlation of expression profiles and between-class highly distinguishable patterns.

The work of Staunton *et al.* (2001) is an example of a supervised learning approach. Specifically, the authors investigated whether patterns of gene expression were sufficient to predict sensitivity or resistance of the NCI 60 cell lines to 232 chemical compounds whose GI activity profile had been previously measured. They measure gene expression of 6,817 genes in each of the 60 untreated cell lines using Affymetrix chips. Chemosensitivity prediction was modeled as a binary classification problem, and thus for each compound two classes of cell lines were defined: sensitive (class 1) and resistant (class 2), according to the GI profiles. They then divided the dataset into a training set and a test set. The classifier was implemented using a weighted voting algorithm, in which correlated genes "vote" on whether a cell is predicted to be sensitive or resistant. Correlation in the training set between a compound c and a gene g is defined as:

$$p(g,c) = \frac{\mu_1(g) - \mu_2(g)}{\sigma_1(d) + \sigma_2(d)} \tag{2.2}$$

Large values of the correlation $P(g,c)$ indicate that the gene expression is a good indicator of class distinction. A weighted sum of the gene expression level of strongly correlated genes is then used to classify. Classifiers with up to 200 genes were tested, with the median accuracy of the classifiers reaching 75%. From this work one can conclude that indeed gene expression profiles in untreated cells can be used to predict whether a cell line is sensitive or resistant to a particular drug.

Other interesting examples of supervised classification methods applied to drug-treated human neural cell cultures come from two studies of Gunther and colleagues. The aim of first study (Gunther *et al.*, 2003) was to investigate whether high-content statistical categorization of drug-induced gene expression profiles can be used to predict the drug's therapeutical class among different classes of psychoactive compounds. Primary cultures of human neuronal precursor cells were treated with multiple members of antidepressants (ADs), antipsychotics (APs), and opioid receptor agonists (OPs). Arguably, however, one of the most used class of psychoactive drugs, the class of antianxiety compounds, would have been an interesting choice. Gene expression was measured using DNA microarrays containing about 11K oligonucleotide probes. Data were analyzed by supervised statistical classification including classification tree (CT) and random forest (RF) methods. Both methods are based on a "leave-one-out" training and testing series, so that the class of the naive test sample can be predicted after training over all other samples. The former method resulted in 88.9% of correct predictions, and relied on few strong markers. Notably, accuracy did not decline significantly when the classification was repeated after withholding the predominant classifier genes from the analysis. The latter method is based on stochastic feature evaluation, and resulted in a correct prediction rate of 83.3% based on a much larger set (326) of week marker genes. Interestingly, two examples are given in which one subclass of AD (SSRIs, or tricyclic) could be successfully predicted as belonging to the antidepressants class after being excluded from the training using the RF. Although the accuracy of prediction of novel subclass unrepresented in the training was surprisingly high (100%), it is unclear why a similar analysis after withholding the third subclass of AD adopted in this study, the MAOIs, is missing. The authors of this work recently published a new study (Gunther *et al.*, 2005) in which they propose a novel algorithm for drug efficacy-profiling, called sampling over gene space (SOGS), and applied it to drug-treated human cortical neuron 1A cell line. While less appealing from a physiological point of view, cell line monocultures provide a simpler system more suitable for reproducible chemical genomics screening. This procedure is based on supervised classification methods such as linear discriminant analysis (LDA) and support vector machines (SVMs), expected to yield stronger predictions than stochastic feature evaluation such as RFs, on one hand, but on the other more prone to "overfitting" the training data. SOGS, however, builds multiple classifier methods iteratively sampling random sets of features using LDA or SVM, and the final classification is based on the most frequent classification over the multiple iterations. The authors claim that such a combination of stochastic feature evaluation with the stable LDA and SVM modeling methods minimizes overfit, while increasing prediction strength.

Gene network reconstruction

Perturbations to the state of the cell have been used extensively in molecular biology to infer the function of a single gene or protein. With the advent of high-throughput quantitative methods it has become possible to move from a qualitative biology to a quantitative biology, thus enabling the use of methodologies typical of engineering and physics to the study of the biological processes and the emergence of "systems biology," i.e., the integrated study of biological processes [for a good review of systems biology refer to Brent (2004) and for its application in drug discovery refer to Butcher *et al.* (2004) and Apic *et al.* (2005)]. Biological processes are the result of complex interaction among thousands of components. Network or graph theory is a mathematical formalism that is very well suited for describing such interactions. Hence the renewed interest in network theory and its potential impact on molecular biology and medicine.

In the area of drug development, particular relevance assumes "reverse engineering" whose goal is to map gene, protein and metabolite interactions in the cell, thus elucidating the regulatory circuits used by the cell for its functioning, and their malfunctioning during diseases. A very good review was recently published on this topic (Gardner and Faith, 2005).

We can distinguish two different reverse engineering strategies (Gardner and Faith, 2005): the "physical approach" and the "influence approach." In the former, the aim is to use RNA expression data to identify the transcription factors (TFs) and the DNA binding sites to which the factor binds. The interactions thus inferred are true physical interactions between TFs and the promoters of the regulated genes. In the latter, the aim is to find regulatory influences between RNA transcripts that do not necessarily have to be of the TF-DNA binding site kind. The general model, as shown in Fig. 2.2, requires that some RNA transcripts act as regulatory "inputs" whose concentration variations drive the expression of an "output" transcript. Such a model therefore does not describe physical interactions, since an mRNA does not control directly the level of other mRNAs, but rather aims at inferring the regulatory influence between two or more transcripts that may as well be indirect through the action of proteins, metabolites and other molecules.

Reverse engineering algorithms make use of measurements of transcript concentrations in response to perturbations to the state of the cell in order to infer regulatory interactions.

In what follows, we will describe examples of the three most successful methodologies for reverse engineering that are currently used.

The first methodology is based on a deterministic approach that describes a gene network as a system of ordinary differential equations (De Jong *et al.*, 2004). The rate of change in concentration of a particular transcript, x_i, is given by a nonlinear influence function, f_i, of the concentrations of other RNAs:

$$\frac{dx_i}{dt} = f_i(x_i, \ldots, x_n) \tag{2.3}$$

where n is the number of genes or transcripts in the network. The function f_i can have different forms. The easiest form that this function can assume is the linear form where Eqn (2.3) becomes:

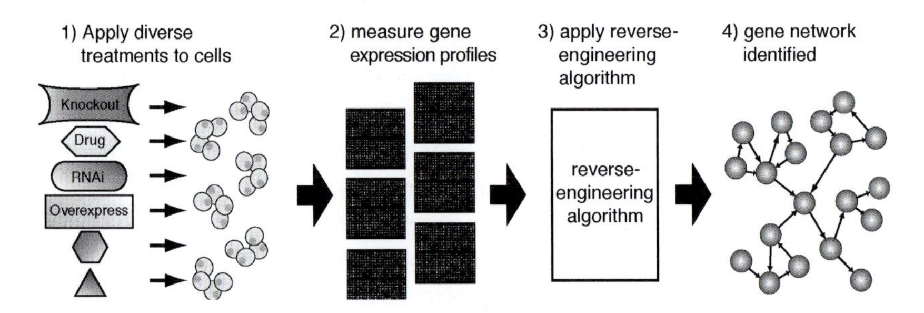

Figure 2.2 Reverse-engineering approaches to drug discovery. Gene expression profiles following a variety of perturbations to the cells are used to reconstruct the network of interactions of gene, proteins and metabolites.

$$\frac{dx_i}{dt} = \sum_j w_{ij} x_i + p_i \qquad (2.4)$$

where w_{ij} represents the influence of gene j on gene i, and p_i an externally applied perturbation to the level of transcript i. We developed an inference algorithm named network identification by regression (NIR) (Gardner *et al.*, 2003) that uses the differential equation model of a gene network in Eqn (2.4) to infer the regulatory interactions among nine genes part of the *Escherichia coli* SOS pathway. The strategy we adopted was to overexpress each of the nine genes in the network using an exogenous plasmid carrying a copy of the gene under the control of an inducible promoter. After transfection and induction of the vector, the gene expression change of the nine genes in the network was measured at steady state, i.e., when the cell has reached a new equilibrium and all the transient effects are over. Under these conditions, the term on the left hand-side of Eqn (2.4) becomes:

$$\frac{dx_i}{dt} = 0$$

so that the equation can be rewritten as:

$$-p_i = \sum_j w_{ij} x_j \qquad (2.5)$$

where both p_i and x_j for all the nine different perturbation experiments are experimentally measured, whereas the weights w_{ij} are the unknown parameters that we would like to learn from the data. Using multiple linear ridge regression, we were able to recover a network model, shown in Fig. 2.3, that correctly identified 25 of the previously known regulatory interactions between the nine transcripts, as well as 14 interactions that could be novel, or possibly false positives. These results were obtained with a noise-to-signal ration of 68%. From a drug discovery point of view, this approach would be powerful for finding new targets for antibiotics, since the nine genes are part of the SOS pathway involved in response to DNA damage. The genes that are the "hubs" of the network, i.e., those genes that are the main regulators of the system, are ideal targets for new antibiotics because they would block the response of the bacteria to damage, thus preventing their survival.

The second methodology is based on an information-theoretic approach that does not use a model to describe gene networks, but rather the information content of gene expression profiles. As an example of successful network inference using this approach applied to a mammalian system, we will illustrate the work of Basso *et al.* (2005). Their approach, named ARACNE (Algorithm for the Reconstruction of Accurate Cellular Networks), is based on the computation of mutual information among pair of genes. For a pair of discrete random variables, $_i$ and $_j$, the mutual information is defined as:

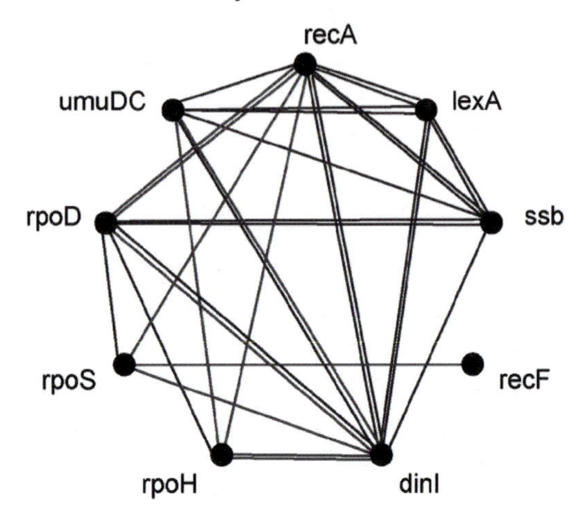

(b) Network model: connection weight matrix

	recA	lexA	ssb	recF	dinI	umuD	rpoD	rpoH	rpoS
recA	**0.40**	**-0.18**	**-0.01**	0	**0.10**	0	**-0.01**	0	0
lexA	**0.39**	**-0.67**	**-0.01**	0	**0.09**	**-0.07**	0	0	0
ssb	**0.04**	**-1.19**	**-0.28**	0	**0.05**	0	**0.03**	0	0
recF	0	0	0	0	0	0	0	0	0
dinI	**0.28**	0	0	0	**-1.09**	**0.16**	**-0.04**	**0.01**	0
umuDC	**0.11**	**-0.40**	**-0.02**	0	**0.20**	**-0.15**	0	0	0
rpoD	**-0.17**	0	**-0.02**	0	**0.03**	0	**-0.51**	**0.02**	0
rpoH	**0.10**	0	0	0	**0.01**	**-0.03**	0	**0.52**	0
rpoS	**0.22**	0	0	**-1.68**	**0.67**	0	**0.08**	0	**-2.92**

Figure 2.3 Inference of a nine-transcript subnetwork of the SOS pathway in *E. coli* using the NIR algorithm. (a) Graph depiction of the network model identified by the NIR algorithm. Previously known regulatory influences are marked in blue, novel influences (or false positives) are marked in red. The strengths and directions of the identified connections are not labeled in the graph. (b) The network model is also depicted as a matrix of interaction strengths. The colors are the same as in (a).

$$I(x,y) = S(x) + S(y) - S(x,y) \tag{2.6}$$

where $S(x)$ defines the entropy. For a given discrete stochastic variable t the entropy is defined as:

$$S(t) = \sum_i \Pr(t = t_i)\log(\Pr(t = t_i)) \tag{2.7}$$

As can be intuitively appreciated from the above definition, entropy is maximal for a uniformly distributed variable. The probability is estimated using Montecarlo simulations. To each value of the mutual information $I(x,y)$ is associated a p-value computed again using Montecarlo simulations. The null hypothesis associated to the p-value corresponds to pair of genes that are disconnected from the network and from each other. The final step of their algorithm is a pruning step that tries to reduce the number of false positives (i.e., inferred interactions among two genes that are not direct interaction in the real biological pathway). They use data processing inequality (DPI) principle that asserts that if both (x,y) and (y,z) are directly interacting, and (x,z) are indirectly interacting through y, then $I(x,z) \leq I(x,y)$ and $I(x,z) \leq I(y,z)$. This condition is necessary but not sufficient, i.e., the inequality can be satisfied even if (x,z) are directly interacting, therefore the authors acknowledge that by applying this pruning step using DPI they may be discarding some direct interactions as well. The authors applied their algorithm on a dataset consisting of 336 whole-genome expression profiles representative of perturbations of B-cell lines and are able to find novel direct targets of the transcription factor MYC.

The third approach is based on Bayesian networks, named after Thomas Bayes (c. 1702 to April 17, 1761), who proved a special case of what is called Bayes' theorem. A Bayesian network is a graphical model for probabilistic relationships among a set of variables. The advantages of using Bayesian networks are (1) it can handle incomplete datasets; (2) it allows one to learn about causal relationships; (3) it can facilitate the combination of domain knowledge and data; (4) it offers an efficient and principled approach for avoiding the over fitting of data; and (5) owing to its probabilistic nature, it is also able to handle noisy data as found in biological experiments.

A Bayesian network describes the relationship between variables (i.e., gene transcript) at both a qualitative and quantitative level. At a qualitative level, the relationships between variables are simply dependence and conditional independence. These relationships are encoded in the structure of a directed graph. At quantitative level, relationships between variables are described by a family of joint probability distributions that are consistent with that independence assertions embedded in the graph.

Thus, to reverse engineer a Bayesian network model of a transcription network, we must find two sets of parameters: the model topology (i.e., the regulators of each transcript), and the conditional probability functions relating the state of the regulators to the state of the transcripts. As in the case of differential equation models, the model-learning algorithm usually presupposes the form of the conditional probability function. Any functions can be used, including Boolean and linear functions. But there will be a tradeoff between model realism and model simplicity. More realistic models will have more parameters, which will require more experimental data and greater computational effort to solve. Bayesian networks can be applied to steady-state data or also to gene expression time-course, in which case the term dynamic Bayesian network is used.

To determine the network structure, usually a heuristic search method is used, like greedy-hill climbing approach or Markov chain Monte Carlo method or simulated annealing. For each network structure S visited in the search, an algorithm learns the maximum likelihood parameters for the conditional distribution functions. It then computes a score for each network S that evaluates how probable it is that the network explains relationships in the observed data D. This is done using either of the two scoring metrics; Bayesian information criteria (BIC) or Bayesian Dirichlet equivalence (BDe). Both scoring metrics incorporate a penalty for complexity to guard against over fitting of data. The BDe score is based on the full Bayesian posterior probability P(S|D) and has an inherent penalty for complexity since it computes the marginal likelihood $P(D|S)$ by integrating the probability of the data over all possible parameters assignments to S. The BIC score is an asymptotic approximation to the BDe score that used an explicitly penalized estimate to the likelihood. One then selects the highest-scoring network as the correct network.

The problem with the Bayesian network is that in the absence of complete dataset the learning problem is underdetermined and several high-scoring networks are found. To address this problem, one can use model averaging or bootstrapping to select the most probable regulatory interactions and to obtain confidence estimates for the interactions. For example, if a particular interaction between two transcripts repeatedly occurs in high-scoring models, one gains confidence that this edge is a true dependency. Alternatively, one can augment an incomplete dataset with the prior information to help select the most likely model structure. The probabilistic structure of a Bayesian network enables straightforward incorporation of prior information via applications of Bayes' rule. The second problem with the Bayesian network is that it cannot contain cycles (i.e., no feedback loops). This restriction is the principal limitation of static Bayesian network models, but dynamic Bayesian network can overcome this limitation.

For an easy-to-use and open-source algorithm to infer gene networks from microarray data using the Bayesian network approach we refer the reader to the Banjo website (http://www.cs.duke.edu/~amink/software/banjo/), a free software suite developed by Alex Hartemink of Duke University (USA).

Recently Luedi *et al* (2005) described an extension of the Bayesian network inference algorithm by which is possible to build a network of biological regulation utilizing biological information by different kind of experimental analysis: it is a method for jointly learning dynamic models of transcriptional regulatory networks from gene expression data and transcription factor binding location data.

Data collected using different technologies offer different perspectives on a problem: joint analysis is likely to produce more accurate results since noise characteristics and biases of the various technologies should be largely independent. In their work, it is shown that supplementing expression data with location data are useful both in increasing the accuracy of recovered networks, and in reducing the quantity of expression data needed to achieve accuracy comparable to that of expression data alone.

As an example of this approach, they applied their joint algorithm to uncover networks describing the regulation of transcription during the cell cycle in yeast, using publicly available cell cycle gene expression data and transcription factor binding location data. The gene expression data consist of 69 time points collected over eight cell cycles. Because these data belong to different phases they use only three states for the phase variable, corresponding roughly to G1, S + G2, and G2 + M. For each of the four synchronization protocols in the dataset (alpha, cdc15, cdc28, and elu) they constructed a phase label. The DBN inference algorithm was applied to a set of 25 genes, 10 of which are known transcription factors with the annotation about their location data (the remaining 15 genes in their set are selected on the basis of the known regulation by one or more of these 10 transcription factors). As an evaluation criterion they created a "gold standard" network consisting of the set of edges that are known to exist from one of the 10 transcription factors with both expression and location data to any one of the other 25 genes in the set.

Results show that the binding site location data by itself does noticeably better than the expression data; despite the relatively poor performance of the expression data when considered in isolation, when it is used with the new informative prior to include evidence from the expression data along with the location data, the number of false-positives and the number of false-negatives are both reduced.

Target validation, hit identification, lead identification, and optimization

Network identification can be used to infer the direct gene and protein targets of a compound with unknown mode of action. One of the earliest approaches of this kind has been proposed by Imoto *et al.* (2003). Although the approach described in the paper is somewhat confusing, we decided to include it in our review since to our knowledge this is one of the first papers to propose that network inference can be used for lead optimization. The authors termed their approach the "virtual gene technique." Briefly, using an algorithm by Maki *et al.* (2001) they reconstruct a directed acyclic graph (DAG) describing gene regulatory interactions considering the drug as a "virtual gene." Let $V = \{g_1, g_2, \ldots, g_n\}$ be the set of all genes and $D = \{d_1, d_2, \ldots, d_n\} \subseteq V$ be the set of genes to be knocked out in order to perturb the system. D is assumed to contain also the virtual gene and the perturbation experiment associated to this virtual gene is treatment with the drug. By observing how the genes change in response to the gene disruption they are able to find a DAG by drawing an edge between two nodes of the graph if a certain equivalence relationship is satisfied. By considering the DAG whose root is the virtual gene, the children of this virtual gene would be the candidate genes directly affected by the drug. From their paper is not clear how well their method performs since the experimental results on deletion strains of *S. cerevisiae* are poorly described. However, their method is an illustrative example of how network inference can be applied to drug discovery.

Another example of network inference to drug discovery is the work of Haggarty *et al.* (2003). Their approach is based on the wild-type and nine different gene deletion strains in *S. cerevisiae*. Each of the strains is treated with all the possible combinations of two molecules drawn from a set of 24 small molecules. The authors propose a method that can be used to understand which of the molecules have similar mode of action by measuring the similarity of chemogenomic networks. For each strain the data were represented as an adjacency matrix, A, with one row and one column for each of the 24 molecules tested. The element a_{ij} of matrix A is 0 when no observable effect on growth after treatment with compound i and j is found, and 1 if there is a measurable growth defect. For each compound in each strain, information in A can be used for clustering the compounds on the basis of similarity in their pattern of biological activity. However, the authors do not test this prediction thoroughly.

The NIR algorithm we developed and briefly described in the previous section can also be used for compound mode of action discovery. The network model can be used as a predictive tool for analyzing new RNA expression data obtained by measuring transcript responses to a drug treatment. As a proof-of-principle, we applied the antibiotic mitomycin C to *E. coli*, and we observed the changes in all nine measured SOS transcripts. However, the known mediator of mitomycin C is only the gene *recA*. The network model obtained by the NIR algorithm enables us to separate secondary changes from primary changes due to direct interaction with the drug. In this case Eqn (2.5) can be solved to find the p_i value for each $i = 1 \ldots 9$, since the network model w_{ij} is known while x_j are the measured response of the cell to the drug treatment. If p_i is close to 0, then gene i is not a direct target of the drug, otherwise gene i is directly interacting with the compound.

The network model correctly filters the RNA expression response to the drug treatment to reveal the recA gene as the direct target. The same target was identified for treatment with UV irradiation and the antibiotic pefloxacin, both of which stimulate *recA* transcript, but not for novobiocin, a drug that should not directly interact via the *recA* gene.

We recently proposed an extension of the NIR algorithm called mode of action by network identification (MNI), that computes the likelihood that gene products and associated pathways are targets of a compound (di Bernardo *et al.*, 2005). Our approach is described in Fig. 2.4. We first reverse engineer a network model of regulatory interactions in the organism of interest using a training dataset of whole-genome expression profiles. The network model is based on ordinary linear differential equations under steady-state conditions described by Eqn (2.5). We then use the model to analyze the expression profile of the compound-treated cells to determine the pathways and genes targeted

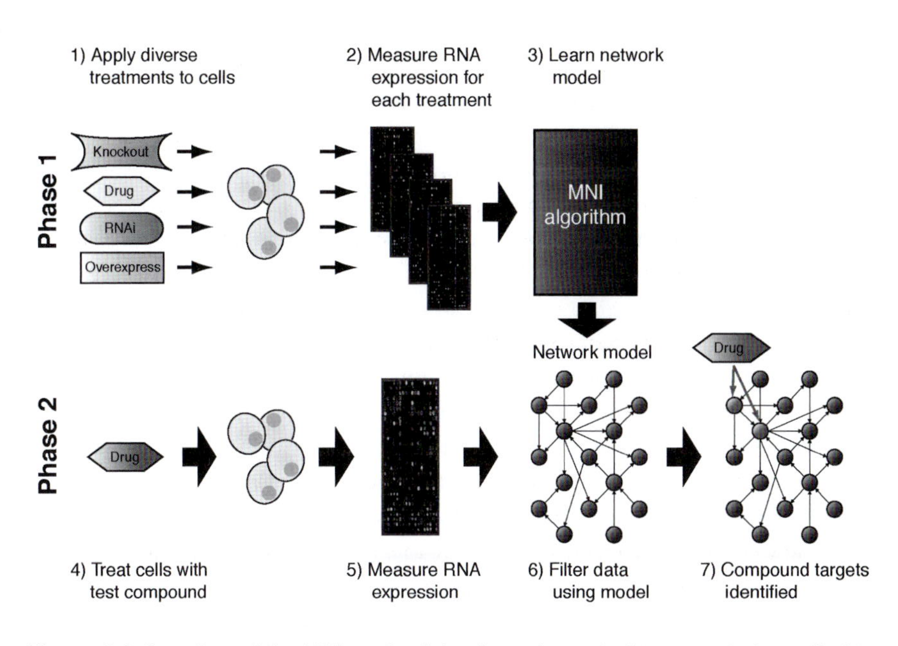

Figure 2.4 Overview of the NMI method. In phase 1, a set of treatments is applied to cells. Changes in mRNA species are measured. The data are then used by the MNI algorithm of infer a model of the regulatory network among the genes. In phase 2, cells are treated with the test compounds and the expression changes of all the mRNA species is measured. The expression data are then filtered using the network model to distinguish the targets of the test compound from secondary responders.

by the compound. The algorithm assumes that the expression profile training dataset are obtained at steady-state following a variety of treatment, including compounds, RNAi, and gene-specific mutations (Fig. 2.4).

The ability to use different treatment types is an important advance over earlier model estimation techniques that require knowledge of the targets of the perturbations. To infer a network model without requiring gene-specific perturbations the algorithm employs an iterative procedure. It first predicts the targets of treatment using an assumed network model, and then uses those predicted targets to estimate a better model. The procedure stops once convergence criteria are met. Once the regulatory model has been learned, we applied it to the expression profile of a test compound to predict its targets. We applied this method to the *S. cerevisiae* using as a training dataset 515 whole-genome yeast expression profiles resulting from a variety of treatment (Hughes *et al.*, 2000; Mnaimneh *et al.*, 2004). We then used MNI algorithm to identify the probable targets of 15 compounds, 13 of which were drawn from the Hughes compendium (Hughes *et al.*, 2000) and from other studies (Ueda *et al.*, 2003). Of these 15 compounds, nine had previously known targets, while the targets of other six were previously unknown. MNI ranks the ~ 6,000 genes in yeast according to their probability of being direct targets of the compound. By selecting the top 50 genes predicted by MNI for a compound, it is possible to infer the pathways directly affected by the drug looking for significantly overrepresented gene ontology (GO) processes among the highly ranked genes.

For seven out of nine compounds with known mode of action, MNI correctly identified the known target pathway and for 6 out of this 9 it was able also to identify the correct target gene. We then demonstrated the use of MNI on a tetrazole-containing compound, 1-phenyl-1H-tetrazole-5-ylsulfonyl-butanenitrile (PTSB) found to inhibit both wt *S. cerevisiae* and human small lung carcinoma cells. We applied MNI to the expression profile after treatment with PTSB and found two genes thioredoxin reductase (TRR1, MNI_rank = 32) and thioredoxin (TRX2, MNI_rank = 36) while the overrepresented GO process among the top 50 genes was the "cell redox homeostasis." We validated the prediction made by MNI with appropriate biochemical assays and confirmed that PTSB acts by inhibiting these two targets.

The method that was discussed before uses microarrays to measure gene expression levels at steady state (i.e., when the cell has reached an equilibrium following a perturbation). This method and the like are quite efficient in recovering the network when it is possible to get the data from diverse experimental conditions, but it misses the dynamic events that are critical for correctly inferring the control structure of a transcription network. At this time in our lab we are implementing an extension of the MNI algorithm that we named time series network identification (TSNI) that works on time series DNA microarray data from cell cultures following a drug treatment (Fig. 2.5). This innovative approach attempts to infer the gene targets and the biological pathways directly affected by a drug by analyzing the gene expression profiles.

The novelty of this approach is in the idea of a gene-centric inference method that can be applied to infer the regulatory interactions of a gene of interest. To this end, we need to measure gene expression profiles at multiple time points following the gene of interests or interest.

The approach used in TSNI is the deterministic approach (Eqn 2.4), similar to that in NIR and MNI. To solve Eqn (2.4), first we need to estimate the rate

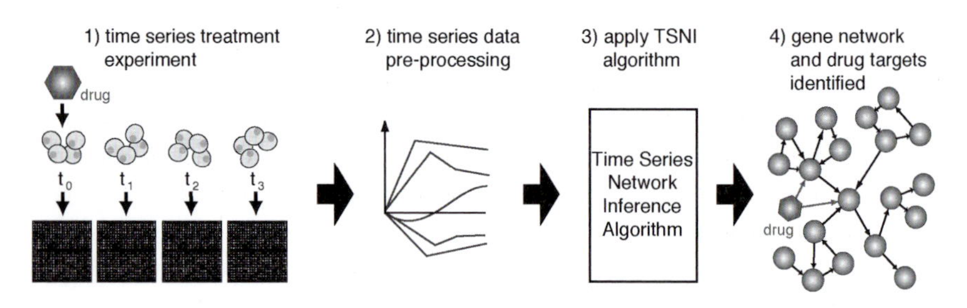

Figure 2.5 The experimental and computational approach to infer compound mode of action from time-course gene expression profiles following treatment of the cells with the drug of interest.

of change of the transcript (dx/dt) from the series. This is problematic because calculating the derivative can amplify the measurement errors in the data. So, we convert the model to a discrete dynamical system, which does not require calculation of the derivatives. Secondly, in general we have a lot more parameters to recover than the number of available data points in the time series, so our problem is underdetermined. To overcome this problem we use principal component analysis (PCA) to reduce the dimension of the data. Once the model is inferred after following the first two steps, we transform it back to its continuous time domain using bilinear transformation, as we know that the solution belongs to the continuous and not the discrete domain.

Our preliminary results (Bansal *et al.*, 2006) based on simulations and on *E. coli* expression data suggest that it is possible to identify the direct targets of the drug. We tested the approach by treating *E. coli* cell lines with norfloxacin. Norfloxacin is a member of fluoroquinolone class of antimicrobial agents that target the prokaryotic type II topoisomerase type II (DNA gyrase) and topoisomerase IV inducing the formation of single-stranded DNA and thus activating the SOS pathway via activation of the recA protein.

We measured six time points with a sampling interval of 6 min after treatment of *the E. coli* cells with norfloxacin (http://m3d.bu.edu). The algorithm analyzed the expression profiles of about ~300 genes that were responding to the perturbation and ranked them according to the probability of being direct targets. Among the top 30 genes, the most overrepresented GO process was the "DNA damage response—SOS pathway," and among the genes in SOS pathway, *recA* was selected the top one. Although preliminary, these results are encouraging and if confirmed would be an important novel tool to determine compound mode of action also in mammalian cells.

Discussion

Computational biology and bioinformatics approaches have the potential to completely change the way drugs are discovered and designed. Already these methods are having an impact on the different stages of the drug discovery process. We have shown in this review how computational methods such as classification and network-based algorithms can be used to understand the mode of action and the efficacy of a given compound and to help elucidating the pathophysiology of a disease. But these computational tools, in our opinion, may also be used in a different and innovative way to promote a change of paradigm in how drugs are designed. In the pharmacological industry there has already been a shift from symptom-oriented drugs, which can relieve the symptoms but not the cause of the disease, to pathology-based drugs, whose targets are the genes and proteins involved in the etiology of the disease. Drugs targeting the affected pathway have thus the potential to become therapeutic. An example of this is enzyme replacement therapy in genetic sulfatase deficiency syndromes (Cosma *et al.*, 2003). The sulfatase enzyme is the missing protein that when reintroduced in the organism is able to restore the pathway that had been altered by the disease process. The passage from symptomatic-centered drug discovery to disease-centered drug discovery has been forced upon the industry by the availability of the full sequence of the human genome, with its implicit promise of novel potential targets. However, as reported in a recent review by Csermely *et al.* (2005), the number of successful drugs did not increase appreciably in the recent years. With the current paradigm, an ideal drug is both potent and specific, i.e., it targets specifically a single protein. In our opinion a second shift is now necessary and will be driven by the availability of sophisticated computational biology and bioinformatics tools: a shift from single-target drugs to "network drugs." By network drug we define a compound or a set of compounds that is able to alter a biological pathway dysregulated by a disease in a predefined way so as to restore its normal physiological function. A similar concept has been put forward by Csermely *et al.* (2005) in their review, in which they propose the partial inactivation of multiple targets as a novel paradigm for drug design. They argue that such kind of multitarget drug could be much more efficient than a drug directed at a single target. They proposed that a network approach to drug design would examine the effect of drugs in the contest of a network of relevant protein-protein, regulatory and metabolic interactions. The end result would be the development of a drug that would hit multiple targets selected in such a way as to decrease network integrity and so completely disrupt the functioning of the network. Our idea is to take this approach one step further and aim not to disrupt the network, but to develop compounds and delivery techniques that are able to change the behavior of the network in a controllable and predictable manner.

Thanks to network inference approaches, some of which were described in this review, it is now becoming possible to have a detailed map of the regulatory circuit among genes, proteins and metabolites. This in turn allows a better understanding of how biological pathways are regulated and how they accomplish their function. The approaches presented in this review also allow the screening of a compound to quickly identify the proteins it interacts with. This gives us all the necessary tools to identify and repair the dysregulated biological pathway causing the disease, much as an engineer would do to restore a malfunctioning electronic circuit. If she/he finds that a specific component of the circuit is malfunctioning, it would be bypassed using extra wires that would bridge different parts of the circuit. Sometimes this would not be sufficient since those parts of the circuit should be in contact only under precisely defined conditions. In this case, she/he would also need to add a microchip that would take care of activating those connections only when necessary.

Similarly, one could think of delivering multiple compounds, each directed to a specific biological target, in a coordinated way controlled by a computer chip that would release the drugs in the organism only when needed to restore physiological behavior of the pathway dysregulated by the disease. The key step in this approach is to have a detailed knowledge of the network of protein, gene and metabolite interactions in the different biological pathways.

Although this picture may seem farfetched, all the tools to accomplish this feat have already been developed and are here to stay, and hopefully in the next decades the way we think of drugs will be completely different.

References

Apic, G., Ignjatovic, T., Boyer, S., and Russell, R.B. (2005). Illuminating drug discovery with biological pathways. FEBS Lett. *579*, 1872–1877.

Baetz, K., McHardy, L., Gable, K., Tarling, T., Reberioux, D., Bryan, J., Andersen, R.J., Dunn, T., Hieter, P., and Roberge, M. (2004). Yeast genome-wide drug-induced haploinsufficiency screen to determine drug mode of action. Proc. Natl. Acad. Sci. USA *101*, 4525–4530.

Bansal, M., Pella Gatta, G., and di Bernardo, D. (2006). Inference of gene regulatory networks and compound mode of action from time course gene expression profiles. Bioinformatics *22*, 815–822.

Bao, L., Guo, T., and Sun, Z. (2002). Mining functional relationships in feature subspaces from gene expression profiles and drug activity profiles. FEBS Lett. *516*, 113–118.

Basso, K., Margolin, A.A., Stolovitzky, G., Klein, U., Dalla-Favera, R., and Califano, A. (2005). Reverse engineering of regulatory networks in human B-cells. Nat. Genet. *37*, 382–390.

Betts, J.C., McLaren, A., Lennon, M.G., Kelly, F.M., Lukey, P.T., Blakemore, S.J., and Duncan, K. (2003). Signature gene expression profiles discriminate between isoniazid-, thiolactomycin-, and triclosan-treated Mycobacterium tuberculosis. Antimicrob. Agents Chemother. *47*, 2903–2913.

Bredel, M., and Jacoby, E. (2004). Chemogenomics: an emerging strategy for rapid target and drug discovery. Nat. Rev. Genet. *5*, 262–275.

Brent, R. (2004). A partnership between biology and engineering. Nat. Biotechnol. *22*, 1211–1214.

Brown, M. P., Grundy, W. N., Lin, D., Cristianini, N., Sugnet, C. W., Furey, T. S., Ares, M., Jr., and Haussler, D. (2000). Knowledge-based analysis of microarray gene expression data by using support vector machines. Proc. Natl. Acad. Sci. USA *97*, 262–267.

Bugrim, A., Nikolskaya, T., and Nikolsky, Y. (2004). Early prediction of drug metabolism and toxicity: systems biology approach and modeling. Drug Discov Today *9*, 127–135.

Butcher, E.C., Berg, E.L., and Kunkel, E.J. (2004). Systems biology in drug discovery. Nat. Biotechnol. *22*, 1253–1259.

Cosma, M.P., Pepe, S., Annunziata, I., Newbold, R. F., Grompe, M., Parenti, G., and Ballabio, A. (2003). The multiple sulfatase deficiency gene encodes an essential and limiting factor for the activity of sulfatases. Cell *113*, 445–456.

Csermely, P., Agoston, V., and Pongor, S. (2005). The efficiency of multi-target drugs: the network approach might help drug design. Trends Pharmacol. Sci. *26*, 178–182.

Dan, S., Tsunoda, T., Kitahara, O., Yanagawa, R., Zembutsu, H., Katagiri, T., Yamazaki, K., Nakamura, Y., and Yamori, T. (2002). An integrated database of chemosensitivity to 55 anticancer drugs and gene expression profiles of 39 human cancer cell lines. Cancer Res. *62*, 1139–1147.

De Jong, H., Gouze, J. L., Hernandez, C., Page, M., Sari, T., and Geiselmann, J. (2004). Qualitative simulation of genetic regulatory networks using piecewise-linear models. Bull. Math. Biol. *66*, 301–340.

di Bernardo, D., Thompson, M.J., Gardner, T.S., Chobot, S.E., Eastwood, E.L., Wojtovich, A.P., Elliott, S.J., Schaus, S.E., and Collins, J.J. (2005). Chemogenomic profiling on a genome-wide scale using reverse-engineered gene networks. Nat. Biotechnol. *23*, 377–383.

DiMasi, J.A., Hansen, R.W., and Grabowski, H.G. (2003). The price of innovation: new estimates of drug development costs. J. Health Econ. *22*, 151–185.

Eisen, M.B., Spellman, P.T., Brown, P. O., and Botstein, D. (1998). Cluster analysis and display of genome-wide expression patterns. Proc. Natl. Acad. Sci. USA *95*, 14863–14868.

Fagan, R., and Swindells, M. (2000). Bioinformatics, target discovery and the pharmaceutical/biotechnology industry. Curr. Opin. Mol. Ther. *2*, 655–661.

Gardner, T.S., di Bernardo, D., Lorenz, D., and Collins, J.J. (2003). Inferring genetic networks and identifying compound mode of action via expression profiling. Science *301*, 102–105.

Gardner, T.S., and Faith, J.J. (2005). Reverse-engineering transcription control networks. Phys. Life Rev. *2*, 65–88.

Gasch, A.P., Spellman, P.T., Kao, C.M., Carmel-Harel, O., Eisen, M.B., Storz, G., Botstein, D., and Brown, P.O. (2000). Genomic expression programs in the response of yeast cells to environmental changes. Mol. Biol. Cell *11*, 4241–4257.

Giaever, G., Flaherty, P., Kumm, J., Proctor, M., Nislow, C., Jaramillo, D.F., Chu, A.M., Jordan, M.I., Arkin, A.P., and Davis, R.W. (2004). Chemogenomic profiling: identifying the functional interactions of small molecules in yeast. Proc. Natl. Acad. Sci. USA *101*, 793–798.

Giaever, G., Shoemaker, D.D., Jones, T.W., Liang, H., Winzeler, E.A., Astromoff, A., and Davis, R.W. (1999). Genomic profiling of drug sensitivities via induced haploinsufficiency. Nat. Genet. *21*, 278–283.

Gunther, E.C., Stone, D.J., Gerwien, R.W., Bento, P., and Heyes, M.P. (2003). Prediction of clinical drug efficacy by classification of drug-induced genomic expression profiles in vitro. Proc. Natl. Acad. Sci. USA *100*, 9608–9613.

Gunther, E.C., Stone, D.J., Rothberg, J.M., and Gerwien, R.W. (2005). A quantitative genomic expression analysis platform for multiplexed in vitro prediction of drug action. Pharmacogenomics J. *5*, 126–134.

Haggarty, S.J., Clemons, P.A., and Schreiber, S L. (2003). Chemical genomic profiling of biological networks using graph theory and combinations of small molecule perturbations. J. Am. Chem. Soc. *125*, 10543–10545.

Hamadeh, H.K., Bushel, P.R., Jayadev, S., DiSorbo, O., Bennett, L., Li, L., Tennant, R., Stoll, R., Barrett, J.C., Paules, R.S., *et al.* (2002a). Prediction of compound signature using high density gene expression profiling. Toxicol. Sci. *67*, 232–240.

Hamadeh, H.K., Bushel, P.R., Jayadev, S., Martin, K., DiSorbo, O., Sieber, S., Bennett, L., Tennant, R., Stoll, R., Barrett, J.C., *et al.* (2002b). Gene expression analysis reveals chemical-specific profiles. Toxicol. Sci. *67*, 219–231.

Hart, C. P. (2005). Finding the target after screening the phenotype. Drug Discov. Today *10*, 513–519.

Hartigan, J. A. (1975). Clustering Algorithms (New York: John Wiley & Sons).

Hastie, T., Tibshirani, R., and Friedman, J.H. (2001). The Elements of Statistical Learning (New York: Springer).

Hughes, J.D., Estep, P.W., Tavazoie, S., and Church, G.M. (2000). Computational identification of cis-regulatory elements associated with groups of functionally related genes in *Saccharomyces cerevisiae*. J .Mol. Biol. *296*, 1205–1214.

Imoto, S., Savoie, C. J., Aburatani, S., Kim, S., Tashiro, K., Kuhara, S., and Miyano, S. (2003). Use of gene networks for identifying and validating drug targets. J. Bioinform. Comput. Biol. *1*, 459–474.

Li, L., Darden, T.A., Weinberg, C.R., Levine, A.J., and Pedersen, L.G. (2001). Gene assessment and sample classification for gene expression data using a genetic algorithm/k-nearest neighbor method. Comb. Chem. High Throughput Screen. *4*, 727–739.

Luedi, P.P., Hartemink, A. J., and Jirtle, R.L. (2005). Genome-wide prediction of imprinted murine genes. Genome Res. *15*, 875–884.

Lum, P.Y., Armour, C.D., Stepaniants, S.B., Cavet, G., Wolf, M.K., Butler, J.S., Hinshaw, J.C., Garnier, P., Prestwich, G.D., Leonardson, A., *et al.* (2004). Discovering modes of action for therapeutic compounds using a genome-wide screen of yeast heterozygotes. Cell *116*, 121–137.

Maki, Y., Tominaga, D., Okamoto, M., Watanabe, S., and Eguchi, Y. (2001). Development of a system for the inference of large scale genetic networks. Pac. Symp. Biocomput. 446–458.

Marton, M.J., DeRisi, J.L., Bennett, H.A., Iyer, V.R., Meyer, M.R., Roberts, C.J., Stoughton, R., Burchard, J., Slade, D., Dai, H., *et al.* (1998). Drug target validation and identification of secondary drug target effects using DNA microarrays. Nat. Med. *4*, 1293–1301.

Mnaimneh, S., Davierwala, A.P., Haynes, J., Moffat, J., Peng, W.T., Zhang, W., Yang, X., Pootoolal, J., Chua, G., Lopez, A., *et al.* (2004). Exploration of essential gene functions via titratable promoter alleles. Cell *118*, 31–44.

Parsons, A.B., Brost, R.L., Ding, H., Li, Z., Zhang, C., Sheikh, B., Brown, G.W., Kane, P. M., Hughes, T.R., and Boone, C. (2004). Integration of chemical-genetic and genetic interaction data links bioactive compounds to cellular target pathways. Nat. Biotechnol. *22*, 62–69.

Parsons, A.B., Geyer, R., Hughes, T.R., and Boone, C. (2003). Yeast genomics and proteomics in drug discovery and target validation. Prog. Cell Cycle Res. *5*, 159–166.

Paull, K.D., Shoemaker, R.H., Hodes, L., Monks, A., Scudiero, D. A., Rubinstein, L., Plowman, J., and Boyd, M.R. (1989). Display and analysis of patterns of differential activity of drugs against human tumor cell lines: development of mean graph and COMPARE algorithm. J. Natl. Cancer Inst. *81*, 1088–1092.

Perlman, Z.E., Slack, M.D., Feng, Y., Mitchison, T.J., Wu, L.F., and Altschuler, S.J. (2004). Multidimensional drug profiling by automated microscopy. Science *306*, 1194–1198.

Ratti, E., and Trist, D. (2001). The continuing evolution of the drug discovery process in the pharmaceutical industry. Farmaco *56*, 13–19.

Scherf, U., Ross, D. T., Waltham, M., Smith, L.H., Lee, J.K., Tanabe, L., Kohn, K.W., Reinhold, W.C., Myers, T.G., Andrews, D.T., *et al.* (2000). A gene expression database for the molecular pharmacology of cancer. Nat. Genet. *24*, 236–244.

Staunton, J.E., Slonim, D.K., Coller, H.A., Tamayo, P., Angelo, M.J., Park, J., Scherf, U., Lee, J.K., Reinhold, W.O., Weinstein, J.N., *et al.* (2001). Chemosensitivity prediction by transcriptional profiling. Proc. Natl. Acad. Sci. USA *98*, 10787–10792.

Stegmaier, K., Ross, K.N., Colavito, S.A., O'Malley, S., Stockwell, B.R., and Golub, T.R. (2004). Gene expression-based high-throughput screening(GE-HTS) and application to leukemia differentiation. Nat. Genet. *36*, 257–263.

Stoughton, R.B., and Friend, S.H. (2005). How molecular profiling could revolutionize drug discovery. Nat. Rev. Drug Discov. *4*, 345–350.

Szakacs, G., Annereau, J.P., Lababidi, S., Shankavaram, U., Arciello, A., Bussey, K. J., Reinhold, W., Guo, Y., Kruh, G.D., Reimers, M., *et al.* (2004). Predicting drug sensitivity and resistance: profiling ABC transporter genes in cancer cells. Cancer Cell *6*, 129–137.

Ueda, M., Kinoshita, H., Yoshida, T., Kamasawa, N., Osumi, M., and Tanaka, A. (2003). Effect of catalase-specific inhibitor 3-amino-1,2,4-triazole on yeast peroxisomal catalase in vivo. FEMS Microbiol. Lett. *219*, 93–98.

Walker, M. G. (2001). Pharmaceutical target identification by gene expression analysis. Mini. Rev. Med. Chem. *1*, 197–205.

Weinstein, J.N., Myers, T.G., O'Connor, P.M., Friend, S.H., Fornace, A.J., Jr., Kohn, K.W., Fojo, T., Bates, S E., Rubinstein, L.V., Anderson, N.L., *et al.* (1997). An information-intensive approach to the molecular pharmacology of cancer. Science 275, 343–349.

Pathway Analysis of Microarray Data

3

Matteo Pellegrini and Shawn Cokus

Abstract
During the past decade, remarkable new techniques for transcriptional profiling have been developed. These include transcriptional profiling using hybridization microarrays as well as methods to sequence transcribed RNAs. No matter which technology is used, these experiments generate data on thousands of genes across multiple conditions and therefore the analysis of such data is often a daunting task. One of the most promising avenues for interpreting large datasets of expression profiles involves pathway-based analysis. Although pathway analysis of expression data is a relatively new field, many important advances have been made over the past few years. Below we outline some of the most significant developments in this area of research.

Introduction

Pathways are collections of genes and proteins that perform a well-defined biological task. For instance, proteins that work to successively synthesize metabolites within a cell are grouped into metabolic pathways. Similarly, proteins that are involved in the transduction of a signal from the cell membrane to the nucleus are grouped into signal transduction pathways. These pathways have been established through decades of molecular biology research and are collected in a variety of public pathway repositories (Kanehisa *et al.*, 2004; Ashburner *et al.*, 2000).

Since the number of known pathways within cells is significantly smaller than the number of genes that is typically profiled, the transformation of data from a gene-centric view to a pathway-centered one represents a dramatic reduction in the number of dimensions. Such reduction allows a biologist to interpret and understand the data in a manner that is not possible when it is viewed as a collection of individual genes.

Although pathway analysis of expression data is a relatively new field, many important advances have been made over the past few years. Below we outline some of the most significant developments in this area of research. These include analyses that attempt to identify the pathways that are overrepresented among significantly perturbed genes in an experiment along with methods that attempt to identify pathways and networks of molecular interactions directly from expression data. Despite the fact that these analyses will undoubtedly continue to evolve rapidly over the next few years, they have already enhanced our ability to understand the biology that underlies complex experiments.

Term enrichment analysis

A typical analysis of microarray expression data generates a long list of genes that are significant according to some criterion. These may be, for example, genes that are differentially regulated in a ratio experiment, or genes that are significant in an analysis of variance (ANOVA) of groups of samples. No matter how the list is generated, it is usually a daunting task to interpret the underlying biology because these lists tend to contain hundreds of genes. In principle, one could search the literature for each gene in the list to attempt to uncover common relationships among them. However, such an approach would inevitably require many hours of research without guaranteeing that the search was comprehensive.

Several tools have emerged to automate this type of analysis. These programs rely on *a priori* classifications of genes into biological function groups. The Gene Ontology (GO) Consortium (Ashburner *et al.*, 2000) generates one of the most widely used of these classifications. GO terms are related to each other through a directed acyclic graph (DAG). That is, most terms have both parent and child terms. The parent terms are more general and inclusive than the child terms. For instance, the parent term "ribonuclear protein complex" (GO id 0030529) has a child term "ribosome" (GO id 0005840). The ontology is separated at the highest level into three separate graphs that contain terms for biological processes, cellular components, and molecular functions. To date, GO represents one of the most comprehensive collections of pathway annotations.

An example of an application that uses GO to automatically perform term enrichment analysis is the Expression Analysis Systematic Explorer (EASE) (Hosack *et al.*, 2003). This tool measures the overlap between a list of genes with GO biological process categories. The significance of the overlap is calculated using the hypergeometric distribution to estimate the probability of finding at least the observed overlap by chance. As an example, the authors computed the GO terms associated with a gene expression study by Kayo *et al.* (2001) on the influence of aging and caloric restriction to the transcriptional profile of skeletal muscle in rhesus monkey. They find that the terms computed with EASE (mitochondrion and electron transport) matched the terms Kayo *et al.* had found through a manual literature search. However, in contrast to the approximately 200 hours required for the literature search, EASE was able to perform the analysis in a few minutes.

Other applications that perform a similar analysis to EASE include GoMiner (Zeeberg *et al.*, 2003), MAPPFinder (Doniger *et al.*, 2003), FatiGO (Al-Shahrour *et al.*, 2004) and GoSurfer (Zhong *et al.*, 2004). These programs differ in the type of gene identifiers that they recognize, the graphical display of the analysis results, the metric that is used to score the enrichment of terms, and the operating system that they work on.

There are many different identifiers that are used to denote genes: gene symbols, Entrez gene identifiers, Affymetrix probe identifiers, SwissProt identifiers, etc. Translating from one id type to another is often a necessary step before any analysis is performed since different applications recognize different identifiers. For example, GoSurfer recognizes Affymetrix probe ids while GoMiner recognizes HUGO gene names. A universal identifier translation tool would be extremely useful but is currently not available; therefore one must manually construct translation files or only use programs that recognize the particular identifiers that one is using.

The output of term enrichment analysis typically comes in two forms: a list of terms that are enriched and a graph of GO terms in which the terms are color-coded according to their statistical significance. If the enriched terms are just reported as a list, the complex relationships between them are not apparent and one may not realize that the significant terms are actually related within the ontology. In contrast, if the output is displayed as a network, then one immediately sees how the terms are related to each other, but may not immediately realize which ones are the most enriched. It is therefore ideal to present the output in both formats. An example of the network output is presented in Fig. 3.1.

Certain programs may be used directly on the web (e.g., FatiGO) while others must be downloaded and installed. Among the latter, some work only on the Microsoft Windows operating system (e.g., GoSurfer) while others are written in Java and are therefore platform independent (e.g., GoMiner).

Gene set analysis

In the previous section, we discussed using the overlap of significant genes in a microarray analysis with functional groups to identify the groups whose members are overrepresented among these genes. In this case, we only consider genes that are deemed to be significant by some threshold and then ignore the particular numerical values. Although this is often a convenient way to select significant genes, in the process we are effectively converting continuous data (gene log ratios or p-values) to binary data (significant or not) and thus losing information. A variety of methods have been developed recently that consider the actual numerical values of genes when attempting to uncover which pathways are of interest.

In one recently published example, the authors attempted to uncover pathways that are differentially expressed between two patient populations: patients

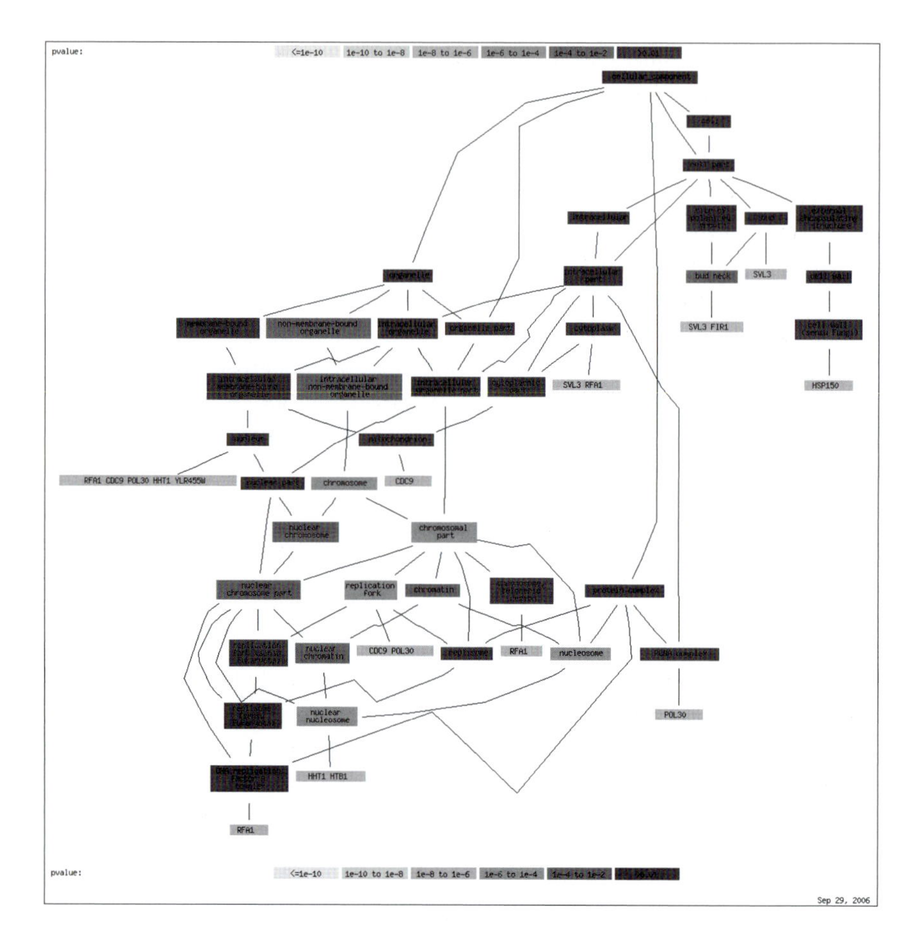

Figure 3.1 The network of enriched GO terms in the cellular component ontology. Different shades of grey indicate that the term is enriched. The figure was generated using the SGD Gene Ontology Term Finder (http://db.yeastgenome.org/cgi-bin/GO/goTermFinder). Genes that are cyclical in synchronized yeast cells were input into the program.

with and without a specific disease (Mootha *et al.*, 2003). The approach used was named Gene Set Enrichment Analysis (GSEA) and attempted to identify the pathway that contained the most differentially expressed genes between the two populations. The analysis was performed on data collected from healthy and diabetic patients. The authors first ranked genes according to the expression difference between the two groups. They then used a Kolmogorov-Smirnov statistic to determine which set of genes had high-ranking members. They were able to estimate the probability of each observation by comparing the real scores to those of randomly permuted data. The analysis identified the oxidative phosphorylation pathway as one of the most differentially expressed and demonstrated that the transcription factor PGC1-α, a regulator of this pathway, and mutations in it correlate with diabetes.

Other approaches similar to GSEA have also been developed. For instance, the program GOMapper computes the significance of the expression of a gene set by computing the ratio of the average expression of genes in the set to the average expression of all genes in the array (Smid *et al.*, 2004). A similar approach is entitled Functional Class Scoring (FCS), wherein the enrichment of each GO term is calculated by estimating the likelihood of observing the product of probabilities of each individual gene associated with the term (Pavlidis *et al.*, 2004). The probabilities associated with each gene are generated from an error model and estimate the likelihood that the gene is perturbed. Monte Carlo simulations estimate the distribution of the products of probabilities to enable the computation of the expectation that a given GO term is enriched for perturbed genes.

The methods described above focus their analysis on datasets in which each gene is assigned a single value. However, these methodologies may be extended to large datasets where gene expression is measured across multiple experiments. In this way, the traditional representation of clustered heat maps of genes versus experiments may be applied to data that measure the activity of pathways across experiments. One example of this type of approach is the map of cancer modules generated by Segal *et al.* (2004). The authors assembled a dataset of 1,975 published microarray experiments that span 22 tumor types. They searched for modules that were significantly active within a subset of experiments. Modules are defined as groups of genes that share a common biological function and are derived by combining multiple sources of gene groupings including GO, KEGG (Kanehisa *et al.*, 2004), and GENMAPP (Doniger *et al.*, 2003).

Among the many conclusions that they could draw from the final module map, they highlight the cell cycle module as active across multiple tumor types, consistent with the observation that all of these tumor types involve rapidly dividing cells. Similarly, many tumor types have an active osteoblastic module, consistent with the fact that many of these tumors metastasize to bone. In contrast, other modules are specific to tumor types. For example, modules that involve neuronal processes are only repressed in a subset of tumors and are otherwise not active. In general, this type of analysis demonstrates that a molecule-level heat map is significantly more interpretable than a gene-level heat map and therefore this approach represents a useful tool for biologists that are trying to cope with large sets of expression microarrays.

Pathway coherence

In order for GO terms or other pathway groupings to appear activated in the previous analyses, the genes within the pathway must be co-regulated. That is, it is unlikely that a random group of genes will ever show up in a pathway analysis, since the genes are independent of each other and unlikely to be perturbed together. In contrast, if a group of genes acts as a single unit (all perturbed together or all unperturbed together) then they are far more likely to appear active. A pathway whose genes are co-regulated may also be called a *coherent pathway*.

It seems reasonable that, if we could determine *a priori* which pathways are coherent and which are not, it might be advantageous to analyze only coherent pathways. One possible metric to measure coherence was developed by Yang *et al.* (2004) and involves measuring the fraction of gene pairs within a pathway that are significantly co-expressed across a set of experiments. Correlation coefficients between pairs of genes are computed and the probability of observing such a correlation or higher is estimated. One may then compute whether the fraction of statistically significant correlations in a pathway group is greater than in a random group of the same size.

Yang *et al.* performed this experiment with normal and tumor tissue samples. They searched for pathways defined by KEGG that had significant coherence. The found that metabolic pathways and protein complexes are coherent while signal transduction pathways are not. This is not surprising since one expects that both metabolic pathways and protein complexes should contain co-regulated genes, while signal transduction pathways on the other hand are controlled by post-translational modifications (e.g., phosphorylation) rather than transcription. A list of the coherent pathways they identified is shown in Table 3.1.

However, not all the genes within a metabolic pathway are co-regulated. Ihmels *et al.* investigated in great detail which components of metabolic pathways are coherent (Ihmels *et al.*, 2004). For example, they found that of the 46 genes assigned to the glycolysis pathway in KEGG, only 24 were correlated in their expression patterns across one thousand diverse experiments. These 24 genes are linearly arranged along the central part of the pathway. They find that in general the central components of metabolic pathways are the most coherent part of the pathway, and that such a central component represents a set of linear reactions.

Ihmels *et al.* also extended their analysis to isozyme pairs contained within metabolic pathways (i.e., pairs of genes with similar sequences that perform slightly different functions). They find that most of the pairs were separately co-regulated with alternative subpathways. In other words, KEGG pathways may often be broken up into distinct subpathways that utilize different members of isozyme pairs. They also identify genes that are co-expressed with the subpathway and are therefore likely components of the pathway. Often such genes code for transporters of the metabolites utilized in the pathway or transcription factors that regulate the pathway.

In summary, it is not only possible to define which pathways are coherent, but also to refine these pathways so that they become more coherent. This involves identifying the most coherent core of the pathway and then extending these cores with additional genes that were not initially associated in the pathway but are co-expressed with the core.

Table 3.1 Coherent pathways

Fructose and mannose metabolism
Sterol biosynthesis
Urea cycle and metabolism of amino acids
Pyrimidine metabolism
Arginine and proline metabolism
Glycoprotein degradation
Ubiquinone biosynthesis
Inositol phosphate metabolism
Sphingoglycolipid metabolism
Nicotinate and nicotinamide metabolism
Apoptosis
Starch and sucrose metabolism
Valine, leucine, and isoleucine degradation
Lysine biosynthesis
Propanoate metabolism
Butanoate metabolism
Protein export
Photosynthesis
Aminoacyl-tRNA biosynthesis
Oxidative phosphorylation
ATP synthesis
Ribosome
Proteasome

Reconstruction of networks using expression data

In the preceding sections, we have discussed techniques for using pre-existing pathway information to interpret microarray expression data. An alternative approach attempts to reconstruct pathways directly from the data. In other words, the previous sections were aimed at supervised analysis whereas here we discuss unsupervised approaches.

One of the first approaches developed to analyze microarray expression data was the clustering program of Eisen *et al.* (1998). They computed pairwise "distances" between genes and clustered genes based on these distances. The approach proved to be remarkably successful in facilitating the interpretation of data. Clusters typically contain genes that function within related pathways or biological processes. It was therefore possible assign functions to previously uncharacterized genes based on the functions of the genes it clusters with.

The reason that pairwise clustering approaches facilitate our interpretation of expression data is that genes with correlated expression tend to function within the same biological process. However, the converse is not often true. That is, genes that are known to function together are not always correlated. In fact, in the majority of cases genes that function together are not significantly correlated. This behavior is consistent with the observation of the previous section that only a minority of pathways are coherent and that only a subset of a typical coherent pathway is in fact truly coherent.

To overcome this difficulty, several methods have been developed to search for functional associations between genes that are not correlated in their expression patterns. One such approach has been to consider higher-order relationships between genes beyond the pairwise ones used in the original clustering methods. For instance, Zhou *et al.* have developed what they term *second-order analysis* (Zhou *et al.*, 2005). Rather than simply calculating pairwise correlations between genes within datasets, the authors compute the correlation between the correlations of two pairs of genes across multiple datasets. In other words, it is possible to first compute the correlation between genes A and B across datasets X and Y: $cAB(X)$, $cAB(Y)$. It is then possible to identify a second pair of genes, C and D, whose correlations in datasets X and Y are correlated with $cAB(X)$ and $cAB(Y)$. The two pairs of genes may have statistically significant second-order correlations even though the pairwise correlations between A and C or B and D are not significant. Thus, it is possible to find relationships between genes that are not captured by pairwise correlations.

To compute second-order relationships, Zhou *et al.* looked at 618 yeast expression arrays that comprised 39 datasets. The analysis revealed 5,142 pairs of genes with significant correlations across some of these datasets and 178,799 statistically significant quadruplets. They observed that 83% of these quadruplets were functionally homogeneous by measuring how often they shared GO terms, implying that the genes participate within the same pathway. In contrast, only 53% of the pairwise relationships were functionally homogeneous. Statistically significant quadruplets seem to group genes into pathways more effectively then pairs. Clustering second-order profiles allows the authors to assign genes to functions more effectively than clustering using Eisen's original approach.

Finally, Zhou *et al.* also apply this approach to transcription factor modules. These are the sets of genes controlled by specific transcription factors. They show that applying second-order analysis allows them to infer regulation motifs in which two transcription factors are being controlled by a third or where one transcription factor is controlling another. They demonstrate that these types of relationships exist between cell cycle transcription factors. For example, the SWI4 and NDD1 modules are correlated in second-order analysis even though none of the genes in one module are correlated with any of the genes in the other. The second-order relationship implies that SWI4 is controlling the transcription of NDD1, and hence the genes regulated by these factors are related in a second-order fashion (Fig. 3.2).

An approach related to second-order analysis and developed by Li *et al.* is named *liquid association* (Li *et al.*, 2004; Li, 2002). The idea underlying this method is that uncorrelated genes may in fact appear correlated when their relationships are conditioned on the state of a third gene. For example, two genes *A* and *B* may appear uncorrelated over a large dataset. However, the pair might appear positively correlated when the values of a third gene C are high and negatively correlated when the values of C are low. This relationship between *A*, *B*, and C may arise if C is somehow controlling the expression of both *A* and *B*.

To illustrate the utility of liquid association, the authors looked at oncogene *P53*. It is known from the literature that *P53* interacts with *TP53INP1* (which encodes a P53-inducible nuclear protein) and *TPBP1* (which codes for P53-binding protein 1). However, these three genes show very low correlation across expression datasets. The authors therefore searched for a fourth gene that possibly interacted with these three and generated a high liquid association score. Their top candidate was *SMARC4*, a gene that encodes a protein that is known to interact with *P53*. The correlations between the three initial genes were there-

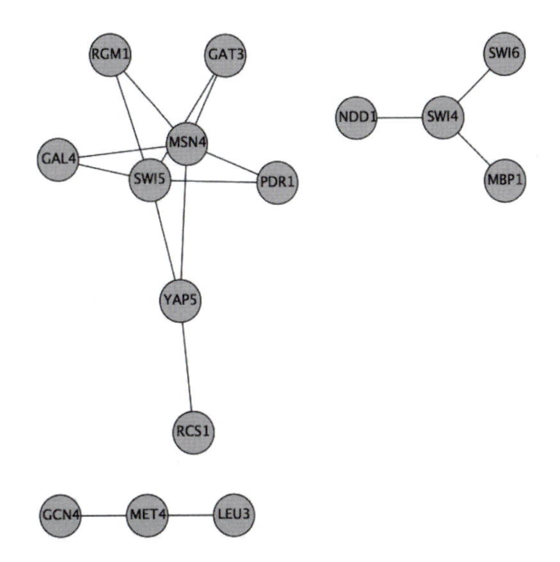

Figure 3.2 Network of second-order interactions between transcription factor modules. Each transcription factor is known to control a group of genes based on transcription factor binding data (Harbison *et al.*, 2004). These gene modules are related by second-order analysis (i.e., the genes from one module are not correlated to the genes in another module, but the correlations between the genes in the two modules are correlated across conditions). These relationships imply that the transcription factors are either interacting or controlling each other, or being controlled by a third factor.

fore significant when conditioned on the expression of *SMARC4*. This may be due to *SMARC4*'s participation in the SWI/SNP transcription factor complex that is necessary for the activation of P53-mediated transcription.

Both second-order analysis and liquid association attempt to identify relationships between small numbers of genes. However, methods have also been developed to reconstruct large networks of gene associations. One such approach utilizes the formalism of Bayesian networks to infer gene networks (Friedman, 2003; Friedman *et al.*, 2000). Bayesian networks model the probability of any state of the system based on the conditional probability distribution of each gene with respect to "parent" genes:

$$P(X_1,\ldots,X_n)=\prod_i P(X_i|U_i) \tag{3.1}$$

where gene expression values are denoted by X_i and corresponding parent genes by U_i. The relationships between genes form a DAG that must be inferred from the data. Inferring the DAG is often computationally expensive. Furthermore, many DAGs provide solutions of roughly the same quality so it is customary to construct an "average network" from all the nearly-optimal inferred DAGs. An example network reconstructed using this technique is shown in Fig. 3.3.

Another approach that has been recently implemented to reverse engineer entire networks is called ARACNE, which stands for Algorithm for the Reconstruction of Accurate Cellular Networks (Basso *et al.*, 2005). This approach uses mutual information to identify pairs of genes that are likely co-regulated. It then applies a filtering step to eliminate pairwise relationships that are likely to be indirect. This filtering step is performed using the "data processing inequality" from data transmission theory. The authors claim that the resulting network is enriched for direct interactions. They also compare this approach to Bayesian networks and demonstrate that in certain cases it yields superior results.

ARACNE was recently applied to the reconstruction of networks in human B-cells. The analysis was performed on 336 B-cell expression arrays that represented a wide collection of normal, transformed, and experimentally manipulated cells. The resulting network includes 129,000 interactions and is therefore difficult to analyze on a global scale. To validate their approach the authors focused their attention on a subnetwork centered around oncogene *MYC*. This network includes 2,063 genes, 56 of which were directly connected to *MYC*. Among the genes that are directly interacting, about half are already known *MYC* targets. They tested 12 of the remaining genes using chromatin immunoprecipitation (ChIP), a technique that allows one to experimentally determine where *MYC* is binding on the genome. They discovered that 11 of these were also *MYC* targets. Therefore, the approach seems to reliably predict which genes are *MYC* targets although its coverage of known *MYC* targets remains sparse.

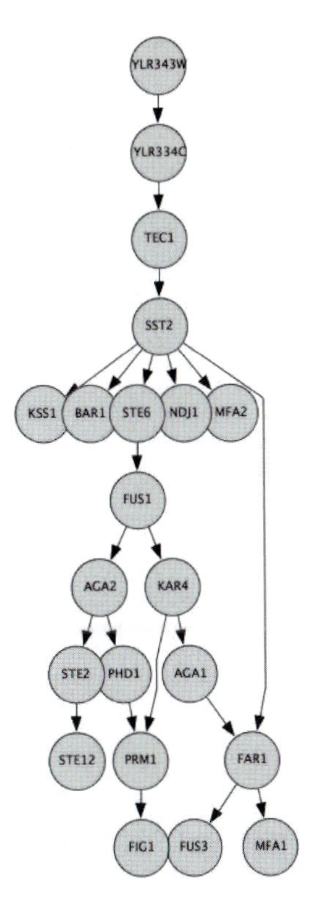

Figure 3.3 Yeast mating Bayesian network constructed by Friedman *et al.* (2003) from a yeast expression dataset of 300 mutant strains. Many of the genes are involved in mating. For instance, we see the mating pheromone a-factor (Mfa1 and Mfa2), along with genes involved in cell fusion (Fus1 and Fus3) and a protease that allows cells to recover from alpha-factor-induced cell cycle arrest by degrading alpha factor (Bar1).

Integrated pathway analysis of expression data and transcription factor binding data

The techniques for pathway analysis of expression data discussed so far have utilized pre-existing pathway information to interpret the data or have attempted to reconstruct networks from expression data. A third class of technique is now emerging that integrates multiple genome-scale data types. In particular, several approaches have been published recently that combine expression data with transcription factor binding data.

ChIP allows one to experimentally determine where transcription factors are binding on the genome. This kind of data has been systematically collected for most of the known transcription factors in *Saccharomyces cerevisiae* (Harbison *et al.*, 2004) and has also been collected less systematically in many other organisms. Over the past few years, a number of techniques have emerged that attempt to interpret expression data in terms of transcription factor binding data and vice-versa.

For example, one of the questions that can be addressed by integrating expression and binding data is in which phases of the cell cycle a transcription factor is active. Alter *et al.* (2004) provided an answer to this question by describing transcription factor binding data in terms of expression data. This technique is inspired by the expression deconvolution technique, in which expression data are represented as a linear combination of basis states (Lu *et al.*, 2003). In Alter's work, the basis states are the components of cell cycle data obtained using singular value decomposition and correspond to the different phases of the cell cycle: G1, S, G2, and M. This analysis allows her to demonstrate that cell cycle transcription factors such as SWI4 and SWI6 are active in the G1 phase whereas the origin replication complex components (such as ORC1) are active in the S phase. In Fig. 3.4, we show that the analysis may be reversed and expression data may be interpreted in terms of transcription factor binding basis states to gauge the activity of each transcription factor within a specific experiment.

Luscombe *et al.* performed a more global analysis of transcription factor activity in yeast (Luscombe *et al.*, 2004). They set out to characterize the transcriptional network across multiple conditions: cell cycle, sporulation, diauxic

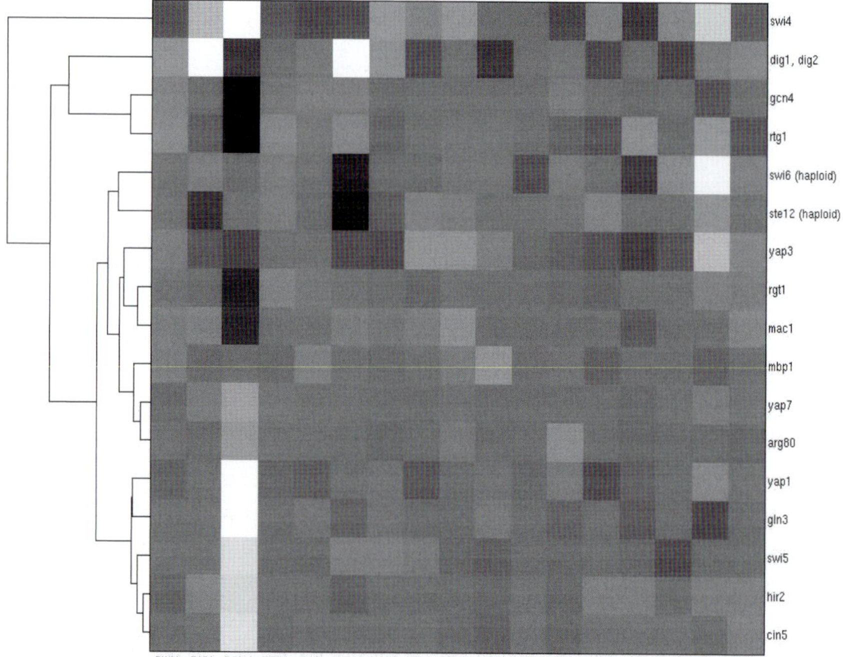

Figure 3.4 A heat map of the activity of transcription factors in yeast deletion experiments. In this example, basis states are defined as the genes bound by a specific transcription factor (the columns). White indicates that the genes bound by the transcription factor are over expressed in the deletion experiments (the rows), while black indicates that the genes are under expressed. This example indicates that expression deconvolution may be used as a proxy for measuring the activity of transcription factors.

shift, DNA damage, and stress response. In each condition they reconstructed a network by identifying transcription factors that were expressed and genes that were differentially expressed and created a link between the two when the binding data suggested the factor bound the gene. They then performed extensive statistical analyses on these networks to identify changes in their properties.

These analyses lead to the classification of the experiments into two broad groups: endogenous processes (cell cycle and sporulation), and exogenous states (diauxic shift, DNA damage, and stress response). The former are complex multistage processes. These have low out degrees (the number of target genes for a given transcription factor), large average pathlengths (the number of links connecting two proteins), and high clustering (the level of transcription factor inter-regulation). In contrast, exogenous states produce rapid, large-scale responses and this is best accomplished with high out degree, small pathlength, and low clustering. In exogenous states, a few transcription factors drive a large number of genes without much "crosstalk."

Conclusions

We have discussed a variety of recent techniques that have been developed to analyze expression data from a pathway perspective. These techniques either leverage existing pathway information or attempt to deduce pathways from the expression data themselves. We have also illustrated how complementary data, such as transcription factor binding, may be used to enhance our understanding of the expression data.

A common perception among biologists is that the interpretation of expression data is one of the primary bottlenecks in the path to scientific discoveries. The pathway analyses we have described are attempts to remove, or at least ameliorate, this bottleneck. They allow scientists to look underneath expression data and interpret what biological phenomena are driving the observed expression patterns. As these techniques mature and become more accessible to the average biologist, expression profiling should become an even more powerful tool than it is already.

Finally, it is important to note that pathway analysis approaches are evolving in parallel with genomic data collection techniques. The availability of new data allows scientists to understand expression data in a deeper manner as we saw in the case of the integration of expression and binding profiles. Since we are merely at the beginning of the technological development of these new profiling techniques, it is reasonable to assume that over the next few years a variety of new pathway analysis approaches will be developed that utilize new types of data. As this occurs, our goal of using expression profiling to transparently interpret the inner workings of the cell will be better realized.

Future developments and trends

The work we have described above provides a static picture of expression data. That is, it allows one to assess which pathways and processes are active in a specific experiment, but not how these change with time. In the future, one may imagine more sophisticated models of expression data that provide such a dynamical view of expression. The advantage of such descriptions would be that one could generate predictions of how the expression of genes would change if experimental conditions are altered.

Detailed dynamical models of biological systems to date have described only small systems that include a few dozen genes and have therefore not been useful in interpreting expression arrays in a general fashion. One exception is the work of Holter *et al.* that generates a simple dynamical model of time series expression data (Holter *et al.*, 2001). Here the authors attempt to derive a model that predicts the state of the system at time t based on its state at an earlier time point:

$$Y(t + \Delta t) = M \cdot Y(t) \tag{3.2}$$

where $Y(t)$ are the expression levels of all genes in an array at time t and M is a "time translation matrix". However, it is not usually possible to solve this equation since the number of time points is typically much smaller than the number of genes. The authors therefore resort to modeling only the primary modes of the time series that are apparent from singular value decomposition. When they apply this approach to model yeast cell cycle data, they demonstrate that the first two modes do a very good job at reproducing the system and that a particular 2×2 time translation matrix M reliably captures the behavior of the system.

Although these results seem promising, this model does not allow one to reliably predict how the system will change in response to different experimental conditions (e.g., mutations or environmental stresses) and therefore the model is still primarily descriptive. Nonetheless, it suggests that future approaches that possibly build upon these types of approaches may in the next few years bring us closer to the realization of truly predictive models. We might then find ourselves in a situation where expression arrays are used to verify models.

References

Al-Shahrour, F., Diaz-Uriarte, R., and Dopazo, J. (2004). FatiGO: a web tool for finding significant associations of Gene Ontology terms with groups of genes. Bioinformatics *20*, 578–580.

Alter, O., and Golub, G.H. (2004) Integrative analysis of genome-scale data by using pseudoinverse projection predicts novel correlation between DNA replication and RNA transcription. Proc. Natl. Acad. Sci. USA *101*, 16577–16582.

Ashburner, M., Ball, C.A., Blake, J.A., Botstein, D., Butler, H., Cherry, J.M., Davis, A.P., Dolinski, K., Dwight, S.S., Eppig, J.T., Harris, M.A., Hill, D.P., Issel-Tarver, L., Kasarskis, A., Lewis, S., Matese, J.C., Richardson, J.E., Ringwald, M., Rubin, G.M., and Sherlock, G. (2000). Gene ontology: tool for the unification of biology. The Gene Ontology Consortium. Nat. Genet. 25, 25–29.

Basso, K., Margolin, A.A., Stolovitzky, G., Klein, U., Dalla-Favera, R., and Califano, A. (2005). Reverse engineering of regulatory networks in human B-cells. Nat. Genet. 37, 382–390.

Doniger, S.W., Salomonis, N., Dahlquist, K.D., Vranizan, K., Lawlor, S.C., and Conklin, B.R. (2003). MAPPFinder: using Gene Ontology and GenMAPP to create a global gene-expression profile from microarray data. Genome Biol. *4*, R7.

Eisen, M.B., Spellman, P.T., Brown, P.O., and Botstein, D. (1998). Cluster analysis and display of genome-wide expression patterns. Proc. Natl. Acad. Sci. USA *95*, 14863–14868.

Friedman, N., Linial, M., Nachman, I., and Pe'er, D. (2000). Using Bayesian networks to analyze expression data. J. Comput. Biol. *7*, 601–620.

Friedman, N. (2003). Probabilistic models for identifying regulation networks. Bioinformatics I, Suppl. 2: II57.

Harbison, C.T., Gordon, D.B., Lee, T.I., Rinaldi, N.J., Macisaac, K.D., Danford, T.W., Hannett, N.M., Tagne, J.B., Reynolds, D.B., Yoo, J., Jennings, E.G., Zeitlinger, J., Pokholok, D.K., Kellis, M., Rolfe, P.A., Takusagawa, K.T., Lander, E.S., Gifford, D.K., Fraenkel, E., and Young, R.A. (2004). Transcriptional regulatory code of a eukaryotic genome. Nature *431*, 99–104.

Holter, N.S., Maritan, A., Cieplak, M., Fedoroff, N.V., and Banavar, J.R. (2001). Dynamic modeling of gene expression data. Proc. Natl. Acad. Sci. USA *98*, 1693–168.

Hosack, D.A., Dennis, G. Jr, Sherman, B.T., Lane, H.C., and Lempicki, R.A. (2003). Identifying biological themes within lists of genes with EASE. Genome Biol. *4*, R70.

Ihmels, J., Levy, R., Barkai, N. (2004) Principles of transcriptional control in the metabolic network of *Saccharomyces cerevisiae*. Nat. Biotechnol. *22*, 86–92.

Kanehisa, M., Goto, S., Kawashima, S., Okuno, Y., and Hattori, M.(2004). The KEGG resources for deciphering the genome. Nucleic Acids Res. *32*, D277–D280.

Kayo, T., Allison, D.B., Weindruch, R., and Prolla, T.A. (2001). Influences of aging and caloric restriction on the transcriptional profile of skeletal muscle from rhesus monkeys. Proc. Natl. Acad. Sci. USA 98, 5093–5098.

Lamb, J., Ramaswamy, S., Ford, H.L., Contreras, B., Martinez, R.V., Kittrell, F.S., Zahnow, C.A., Patterson, N., Golub, T.R., and Ewen, M.E. (2003) A mechanism of cyclin D1 action encoded in the patterns of gene expression in human cancer. Cell *114*, 323–334.

Li, K.C. (2002) Genome-wide coexpression dynamics: theory and application. Proc. Natl. Acad. Sci. USA *99*, 16875–16880.

Li, K.C., Liu, C.T., Sun, W., Yuan, S., and Yu, T. (2004). A system for enhancing genome-wide coexpression dynamics study. Proc. Natl. Acad. Sci. USA *101*, 15561–15566.

Lu, P., Nakorchevskiy, A., and Marcotte, E.M. (2003). Expression deconvolution: a re-interpretation of DNA microarray data reveals dynamic changes in cell populations. Proc. Natl. Acad. Sci. USA 100, 10370–10375.

Luscombe, N.M., Babu, M.M., Yu, H., Snyder, M., Teichmann, S.A., and Gerstein, M. (2004). Genomic analysis of regulatory network dynamics reveals large topological changes. Nature. 431, 308–212.

Mootha, V.K., Lindgren, C.M., Eriksson, K.F., Subramanian, A., Sihag, S., Lehar, J., Puigserver, P., Carlsson, E., Ridderstrale, M., Laurila, E., Houstis, N., Daly, M.J., Patterson, N., Mesirov, J.P., Golub, T.R., Tamayo, P., Spiegelman, B., Lander, E.S., Hirschhorn, J.N., Altshuler, D., and Groop, L.C. (2003). PGC-1alpha-responsive genes involved in oxidative phosphorylation are coordinately downregulated in human diabetes. Nat. Genet. *34*, 267–273.

Pavlidis, P., Qin, J., Arango, V., Mann, J.J., Sibille, E. (2004). Using the gene ontology for microarray data mining: a comparison of methods and application to age effects in human prefrontal cortex. Neurochem. Res. *29*, 1213–1222.

Segal, E., Friedman, N., Koller, D., Regev, A. (2004). A module map showing conditional activity of expression modules in cancer. Nat. Genet. *36*, 1090–1098.

Smid, M., Dorssers, L.C. (2004) GO-Mapper: functional analysis of gene expression data using the expression level as a score to evaluate Gene Ontology terms. Bioinformatics 20, 2618–25.

Yang, H.H., Hu, Y., Buetow, K.H., Lee, M.P. (2004). A computational approach to measuring coherence of gene expression in pathways. Genomics *84*, 211–217.

Zeeberg, B.R., Feng, W., Wang, G., Wang, M.D., Fojo, A.T., Sunshine, M., Narasimhan, S., Kane, D.W., Reinhold, W.C., Lababidi, S., Bussey, K.J., Riss, J., Barrett, J.C., Weinstein, J.N. (2003) GoMiner: a resource for biological interpretation of genomic and proteomic data. Genome Biol. 4, R28.

Zhong, S., Storch, K.F., Lipan, O., Kao, M.C., Weitz, C.J., and Wong, W.H. (2004). GoSurfer: a graphical interactive tool for comparative analysis of large gene sets in gene ontology. Appl. Bioinformatics 3, 261–264.

Zhou, X.J., Kao, M.C., Huang, H., Wong, A., Nunez-Iglesias, J., Primig, M., Aparicio, O.M., Finch, C.E., Morgan, T.E., and Wong, W.H. (2005). Functional annotation and network reconstruction through cross-platform integration of microarray data. Nat. Biotechnol. 23, 238–243.

Toxicogenomics: Applications of Genomics Technologies for the Study of Toxicity

4

Uwe Koch

Abstract

Toxicogenomics is a scientific subdiscipline that combines toxicology with genomics. Toxicogenomics analyses the activity of a toxin or chemical substance on living tissue based upon a profiling of its known effects on genetic material. Toxicogenomics may also be of use as a preventative measure to predict adverse "side," i.e., toxic, effects, of pharmaceutical drugs on susceptible individuals. This involves using genomic techniques such as gene expression level profiling and single-nucleotide polymorphism analysis of the genetic variation of individuals. These studies are then correlated to adverse toxicological effects in clinical trials so that suitable diagnostic markers (measurable signs) for these adverse effects can be developed.

Introduction

Absorption, distribution, metabolism, excretion, and toxicity (ADME/Tox) properties of a molecule determine whether a compound which binds with high affinity to the chosen target can progress into clinical phase. Prediction or determination of these properties before a compound reaches the clinical stage, and thus preventing failure at the final stage is very important. In this report we focus on gene-expression based approaches in preclinical ADME/Tox. There is a need for better methods, preferably based on molecular structure. On the molecular level, the ADME/Tox of a compound is influenced by a system of transporters, channels, receptors and enzymes which act together. To date, empirical data have been used to build computer models which evaluate real and virtual compounds. Improving models requires the collection of high-throughput data, including global gene expression, proteomics and metabolic profiles together with clinical and phenotypic data.

It is desirable for toxicology to become a mechanistic science which is capable of saving development candidates instead of merely terminating them by characterizing specific toxicities. It should be possible to design toxicology studies to address specific mechanistic concerns and allow risk assessment to position finding in animal models relative to exposed humans. Presently, in addition to high-throughput screening assays for binding to receptors and other proteins of interest, much data generated in at least four "OMICS" areas are related to the interaction of a drug with proteins involved in ADME/Tox (Fig. 4.1). These areas are toxicogenomics, proteomics, metabonomics, transcriptomics, and pharmacogenomics. Proteomics is the large-scale study of proteins, particularly their structures and functions. Metabonomics studies the changes in metabolite profiles as a result of a biological perturbation. Transcriptomics studies the expression level of genes, using techniques capable of sampling tens of thousands of different mRNA molecules at a time. Pharmacogenomics deals with the influence of genetic variation on drug response in patients. It correlates gene expression or single-nucleotide polymorphisms with a drug's efficacy or toxicity.

Toxicity always leads to changes in gene expression and protein concentrations. These changes occur at lower concentrations and earlier in time than the toxic endpoint. Thus genomics and proteomics have the potential to deliver a wealth of data on molecular and cellular responses of any model to a toxic agent. Genomics can be divided into three main areas: DNA sequencing, transcription profiling and proteomics, which deal with three different cellular macromolecules, DNA, RNA and proteins. The basis of the genomic revolution is DNA sequencing which reveals genetic information. Transcription profiling, the study of the response of multiple mRNA species present in a given tissue, reveals the cellular response to a treatment by measuring relative changes in mRNA. Changes in proteins including post-translational modifications are studied by proteomics.

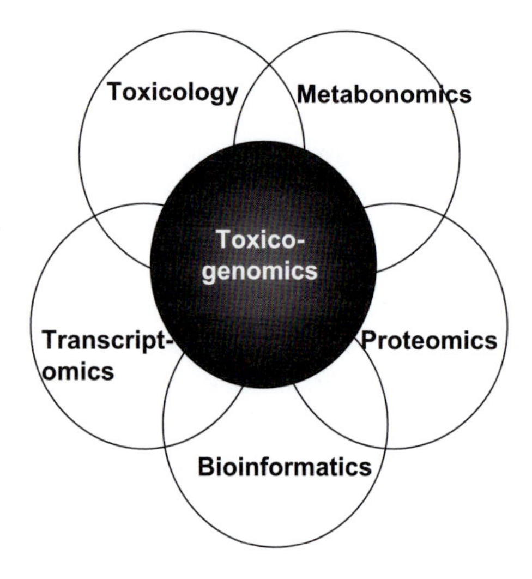

Figure 4.1 Each cycle represents one of the scientific areas contributing to toxicogenomics.

An important objective in toxicogenomics is to integrate data from different studies and analytical platforms to produce a richer and biologically more refined understanding of the toxicological response of a cell, organ or organism. For example, one goal would be to describe the interplay between protein function and gene expression, or between the activity of certain metabolizing enzymes and the excretion of small metabolites. The integration of data from different domains has been reported. In these experiments, tissue samples that were derived from the same individual animals or from comparably treated animals were analyzed in parallel using different technologies. It is essential that the right conclusions are drawn in the multitude of observed relevant and irrelevant effects. Gene expression data need to be put into perspective with other (clinical, pathological) data and an expert evaluation is essential. So, integration is needed of genomics and proteomics into toxicology.

Toxicogenomics in the preclinical testing funnel

The desire for a potent and pharmacologically active molecule is the first step towards a marketed product. The path begins with the development of a small molecule with potency and selectivity for the desired target. Often, a genomic-based approach identifies new targets by analyzing changes of gene expression in disease tissue. After a selective and potent inhibitor has been developed *in vitro* studies can be conducted to characterize the potential for class- and compound-specific toxicity. These experiments are designed to identify fundamental toxicology problems such as genotoxicity, cytochrome P450 induction or over cellular toxicity. These studies are performed in cell lines with little or no characterization of the expression profile of the test system. These assays can be performed quickly with minimal compound expense and represent the first toxicology experiments performed.

Before the first dose in man a large amount of preclinical data must be collected. One goal for the design of the toxicological testing cascade is to identify development-limiting issues as early as possible. However, care must be taken to avoid elimination of good candidates through the use of poorly predictive assays. As in early drug discovery the screening process must be a sequential testing funnel, where certain target criteria are either achieved or not achieved, resulting in removal of the compound from consideration for development (Fig. 4.2). In this process there are numerous opportunities for transcription profiling to contribute as is illustrated in Fig. 4.2. The individual contributions will be discussed in the following chapters starting with mechanistic toxicogenomics.

Mechanistic toxicogenomics

The goal of Mechanistic toxicogenomics (Wang *et al.*, 2003, 2005) is the understanding of a specific toxic effect observed for a compound and is important for the assessment of risk. In the early phase of drug development understanding the mechanisms of toxicity can be important for the decision to continue

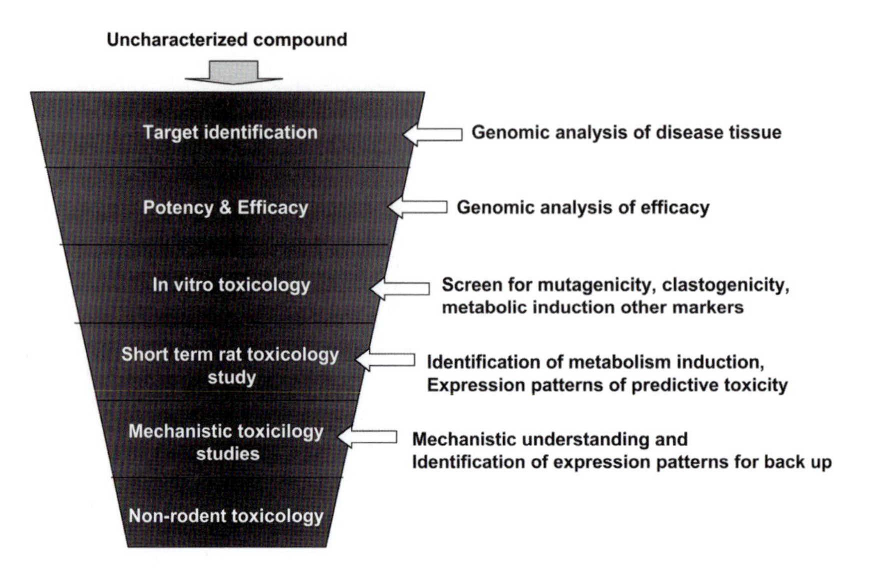

Figure 4.2 The sequential testing funnel describing the tests which a potential drug candidate has to pass and the contribution of toxicogenomics.

development of a compound or to switch to a back-up series. It can also help to decide which series to choose for further development. There are numerous examples in which a specific structural element in a series of compounds has been identified to be responsible for a toxicological problem and its replacement has led to a successful alternative series.

It has been shown that changes in the expression of small gene sets discriminate between mechanistically distinct classes of drugs (Fig. 4.3). These sets of genes can be used to classify compounds and match a compound with an unknown mechanism into a predefined class of compounds with a similar mechanism (Hamadeh *et al.*, 2002a–c; Lord, 2004; Lord and Papoian, 2004; Morgan, 2002).

Phenotypic anchoring

Toxicology uses surrogate markers that are correlated with toxic responses to monitor adverse outcomes in inaccessible tissues. For example, the concentrations of liver enzymes alkaline phosphatase (ALT) and aspartate aminotransferase (AST) correlate with histopathological changes in the liver (Hamadeh *et al.*, 2002c). These serum enzyme markers, together with histopathology, facilitate the "phenotypic anchoring" of molecular expression data (Heinloth *et al.*, 2004; Tennant, 2002). Phenotypic anchoring is the process of determining the relationship between a expression profile and the pharmacological or toxicological phenotype of the organism for a certain exposure or dose and at a particular time. Dose and time alone are often not enough to define the toxicity experienced by an individual animal, so another measure of toxicity is needed for the full interpretation of the data obtained during a toxicogenomics study.

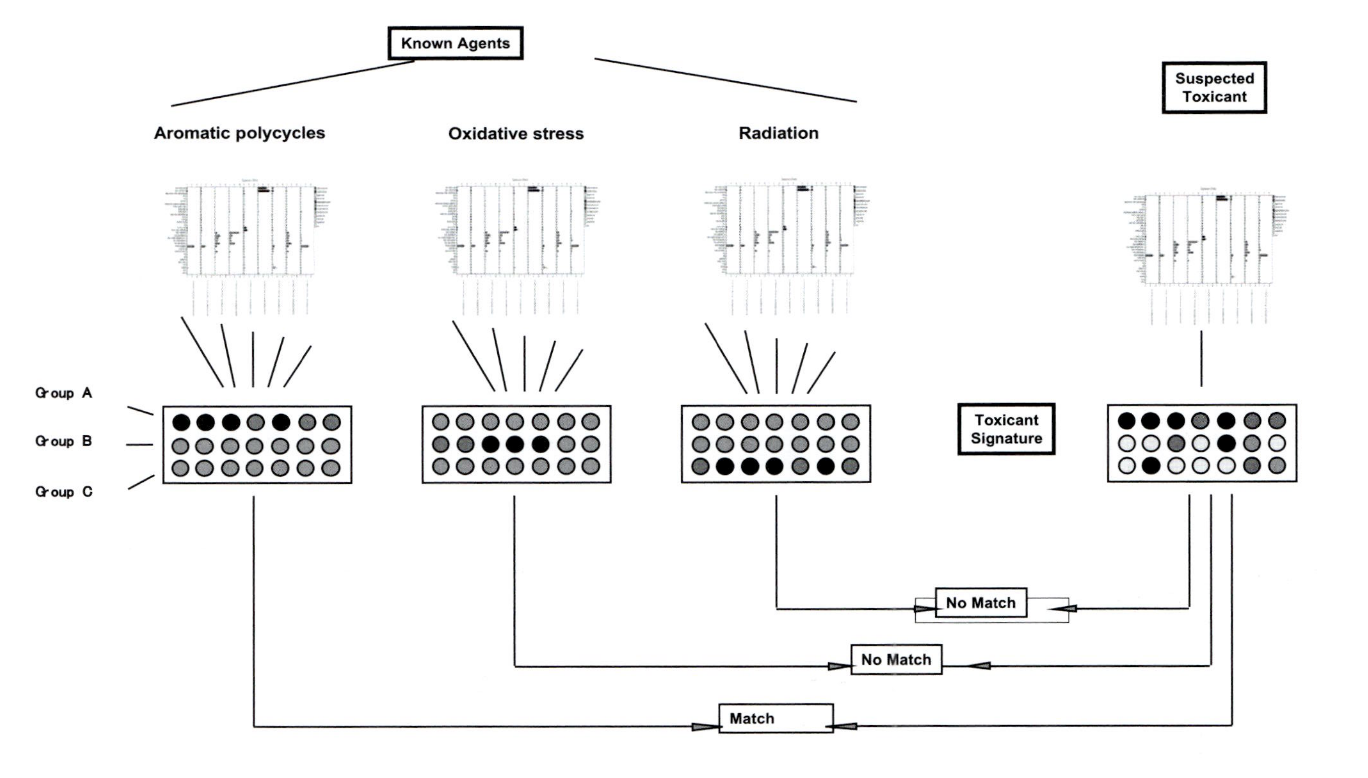

Figure 4.3 It is shown how comparison of the expression profile of a suspected toxicant with the expression profiles of known agents leads to the identification of a expression profile similar to aromatic polycycles.

Conversely, the phenotype alone might be insufficient to anchor the molecular profile, because a high level of ALT in serum can be observed both before peak toxicity (as it rises) and after peak toxicity (as it returns to baseline). For this reason, anchoring the molecular expression profile in phenotype, dose and time serves to define the sequence of key molecular events in the mode-of-action of a toxicant.

Waters *et al.* (2003) have suggested to extend toxicogenomics and combine it with computational approaches such as physiologically-based pharmacokinetic (PBPK) and pharmacodynamic modeling. PBPK modeling can be used to estimate quantitatively the dose of the test agent or its metabolites present in the target tissue at any time after treatment, thereby allowing molecular expression profiles to be anchored to internal dose, as well as to the time of exposure and to the toxicant-induced phenotype. Correlations between gene, protein and metabolite expression can be described as a function of the applied dose of an agent and the kinetic and dynamic dose-response behaviors that occur in different parts of the tissue. These models must take into account the fact that the transcriptome, proteome and metabolome are themselves dynamic systems, and are therefore subject to significant environmental influences, such as time of day and diet (Amin *et al.*, 2002; Kaput and Rodriguez, 2004).

Examples

Testicular toxicity is of particular interest in drug discovery since testicular changes are often subtle in early stages and lack well-established biomarkers or clear morphologic changes. For example, changes in gene expression of mice treated with bromochloroacetic acid, a known testicular toxicant, showed a particularly strong effect on genes involved in fertility, cell communication and adhesion. This supported the hypothesis that toxicologic effect was the result of cellular interactions between Sertoli cells and spermatids (Richburg *et al.*, 2002).

Carcinogenicity is usually evaluated in mouse bioassays that take up to 2 years. Gene expression experiments have been used to identify gene markers speeding up the identification of carcinogens. Rats were treated with several prototype rodent genotoxic and non-genotoxic carcinogens and two non-carcinogenic hepatotoxicants. Correlation of the gene-expression data with the known carcinogenic potential of these compounds has identified transforming growth factor β-stimulated clone 22 and NADP(H) CYP450 oxidoreductase as molecular markers of non-genotoxic carcinogenicity in rodents (Kramer *et al.*, 2004).

Proteomics

Proteomics deals with the measurement of protein concentrations in whole tissue samples and is a valuable complement to gene-expression data (Heijne *et al.*, 2005; Leighton, 2005; Merrick and Madenspacher, 2005). Post-translational modification, proteolysis can change the presence of the active protein so that the mature mRNA does not always correspond to the presence of active protein. Interestingly in humans, there may be only about 30,000 genes, but there are thought to be many more cellular proteins than there are genes, including all the possible post-translational modifications. This poses a theoretical problem when gene expression- proteomic correlations are being sought as there is a higher level of cellular control than the genome which is in the protein body itself. Also changes in gene expression which may or may not result in changes in cellular protein synthesis have to occur at different times in the cell, and different gene regulation events occurring at the same time may take different times to effect the proteome.

Proteomics studies use two-dimensional protein electrophoresis and mass spectrometry to identify changes in the presence of a protein in a tissue under different conditions (Aebersold and Mann, 2003; Lahm and Langen, 2000). As far as toxicity studies are concerned, samples from liver, kidney and plasma have usually been used to evaluate the responses of experimental models to drugs. Proteomics has become an automated procedure in cutting and digesting protein spots from electrophoretic gels and their analysis and identification by mass spectrometry.

Metabonomics

Global metabolic profiling (metabonomics/metabolomics) (Griffin and Bollard, 2004; Keun, 2005; Pognan, 2004; Robertson, 2005; Witkamp, 2005) has shown particular promise in the area of toxicology and drug development. A major advantage of this approach is its high-throughput nature. In recent years there has been a steady development of nuclear magnetic resonance (NMR) to analyze plasma and urine, for instance, to identify changes in levels of natural metabolites in humans and experimental animals. Many of the bioinformatic and statistical techniques employed have much in common with those used in proteomics and toxicogenomics. NMR techniques can be applied to analyze urine samples to detect metabolic responses to pathophysiological stimuli or genetic modification. Changes in levels of individual metabolites from many metabolic systems can be recognized in pattern recognition. This can be used not only for classification but also to identify new metabolites and probes for pathological mechanisms. The general aim of pattern recognition (PR) is to classify objects (in this case ^{1}H-NMR spectra) or to predict the origin of objects based on identification of inherent patterns in a set of indirect measurements. PR methods can reduce the dimensionality of complex datasets via 2D or 3D mapping procedures, thereby facilitating the visualization of inherent patterns in the data. Most work has been done with clinical samples from drugs. A drawback of using NMR is its intrinsically low sensitivity compared to mass spectroscopy for example. Therefore a combined approach to Metabolic Profiling using both methods in a synchronized mode delivers the maximum information content.

Predictive toxicogenomics

The goal of predictive toxicogenomics (Ellinger-Ziegelbauer *et al.*, 2005; Kier *et al.*, 2004; Waring *et al.*, 2002) is to identify potential safety liabilities as early as possible. Predictive toxicology is based on the hypothesis that similar treatments leading to the same end point are accompanied by similar changes in gene expression. A major challenge in predicting toxicologic end points based on transcriptional data lies in discriminating changes due to inter-animal variation or experimental background noise from treatment-related changes. Compounds may directly affect expression of certain well-characterized, compound-specific genes. These compound-specific genes are not suited for discrimination between different classes of compounds since, if related to the pharmacodynamics of the compound, they are common to all compounds of this class. Drugs, in contrast to other toxic substances, have pharmacologic as well as toxicologic effects that might affect gene expression. These two effects can, but need not, be related. Despite these confounding factors, gene expression analysis after treatment with various compounds that result in the same toxicologic end point should enable identification of a toxic fingerprint.

Example

Since gene expression changes occur before histopathological or clinical pathological changes, analysis of gene expression has the potential of discovering compounds with toxic liabilities early in the drug discovery process using 1- or 3-day rat studies. However, it is unlikely that a change in expression of any single gene is strongly correlated with a particular toxicity. Usually it will be a complex pattern of changes in expression, up and down, of many genes which is related to any toxic effect. To identify and classify these patterns statistical tools are employed.

Steiner *et al.* (2004) have shown that compound classification based on gene expression data is feasible. Male rats were treated with various model compounds or the appropriate vehicle controls. Most substances were either well-known hepatotoxicants or showed hepatotoxicity during preclinical testing. The aim was to determine if biological samples from rats treated with various compounds can be classified based on gene expression profiles. In addition to gene expression analysis using microarrays, a complete serum chemistry profile and liver and kidney histopathology were performed. Gene expression profiles were analyzed using a supervised learning method (support vector machines; SVMs) to generate classification rules and to identify a compact subset of probe sets with potential use as biomarkers using feature selection. The predictive models were able to discriminate between hepatotoxic and nonhepatotoxic compounds. Furthermore, they predicted the correct class of hepatotoxicant in most cases. The authors show also that that a predictive model built on transcript profiles from one rat strain can successfully classify profiles from another rat strain. They also demonstrate that the predictive models identify non-responders and are able to discriminate between gene changes related to pharmacology and toxicity.

Data analysis

The experience gained from integrating global-proteomics or metabonomics data, such as spot intensities from 2D gels or metabonomics fingerprint data from NMR, tells that cluster or principal component analysis can be done to derive global signatures of molecular expression. If biological samples segregate into unique clusters that show similar expression characteristics, further efforts can be made to discern the new proteins or metabolites that are expressed in these samples. Further steps can also be taken to evaluate these proteins or metabolites as potential biomarkers and as a means to determine the underlying toxicological response.

Commonly used statistical methods reduce the number of parameters in the classification model. One approach of feature selection is to rank genes according to the differences in expression between experiments. Methods used are: standard or permutation *t*- or *F*-test; ad hoc signal-to-noise statistics; Wilcoxon statistics, or significance analysis of microarrays. (Bushel *et al.*, 2002; Golub *et al.*, 1999; Hastie *et al.*, 2001)

Another approach of feature selection is to use noise reduction methods, such as principal component analysis (PCA) (Khan *et al.*, 2001) and wavelet transformation (Lio, 2003). PCA reduces large sets of genes into components which are linear combinations of all genes. The major information is captured by selecting the first components thus vastly reducing the amount of gene expression data to be analyzed.

After data reduction predictive models can be established which applied to gene expression data from compounds with unknown toxicological profile allow a prediction of potential toxicological problems. A large number of computational algorithms has been developed for this purpose. Examples include naïve Bayesian classifiers, artificial neural networks (ANN) and support vector machines (SVM). Support vector machines (SVMs) belong to the class of supervised learning algorithms. Originally introduced by Vapnik and co-workers (Vapnik and Chapelle, 2000), they perform well in different areas of biological analysis (Scholkopf *et al.*, 2000; Smola *et al.*, 1998). Given a set of training examples, SVMs are able to recognize informative patterns in input data and make generalizations on previously unseen samples. Like other supervised methods, SVMs require prior knowledge of the classification problem, which has to be provided in the form of labeled training data. Used in a growing number of applications, SVMs are particularly well suited for the analysis of microarray expression data because of their ability to handle situations where the number of features (genes) is very large compared with the number of training patterns (microarray replicates). Several studies have shown that SVMs typically tend to outperform other classification techniques (Arimoto *et al.*, 2005; Byvatov *et al.*,

2003; Krishnan and Westhead, 2003; Wang *et al.*, 2005; Wei *et al.*, 2005). In addition, the method proved effective in selecting relevant features such as genes that are especially relevant for the classification and therefore might be critically important for the biological processes under investigation. A significant reduction of the gene number used for classification is also crucial if reliable classifiers are to be obtained from microarray data.

In vitro methods

Most toxicogenomic studies to date have been carried out using tissue from animals dosed *in vivo*. Large amounts of compound are required for these studies which are usually not available in the early phase of the drug discovery process. In addition the number of compounds which can be tested is very limited in animal studies. Thus *in vitro* techniques would significantly increase the value of this technology in drug discovery. As with other *in vitro* systems in drug safety assessment the main question is about their predictive value for toxicity in animals or people. Studies have shown that compounds with different mechanism of toxicity can be distinguished using *in vitro* experiments.

Example

At Abbott the gene expression profiles of 15 hepatotoxins in isolated primary rat hepatocytes was compared (Waring *et al.*, 2003). Using unsupervised hierarchical clustering compounds with different mechanism of toxicity could be clearly distinguished. Using gene expression data the compounds could be correctly classified into different mechanistic hepatotoxin classes. These and other studies demonstrate that gene expression can be used to classify compounds.

Absorption

Here measurement of transporter gene expression is used to predict plasma levels. Membrane transporters and channels (the transportome) govern cellular influx and efflux of ions, nutrients, and drugs. Oligonucleotide arrays have been used to analyze gene expression of the transportome in 60 human cancer cell lines used by the National Cancer Institute for drug screening. Gene expression was correlated with the potencies of standard anticancer drugs and helped to confirm known drug-transporter interactions and suggested novel ones. Folate, nucleoside, and amino acid transporters positively correlated with sensitivity to their respective drug substrates. The positive correlation between the expression of a nucleoside transporter, SLC29A1, and the potency of nucleoside analogues, azacytidine and inosine-glycodialdehyde, was validated. Application of an inhibitor of SLC29A1 significantly reduced the potency of these two drugs, indicating that SLC29A1 plays a role in cellular uptake. Three ABC efflux transporters (ABCB1, ABCC3, and ABCB5) were shown to exhibit significant negative correlations with multiple drugs, suggesting a mechanism of drug resistance. ABCB1 expression correlated negatively with potencies of

19 known ABCB1 substrates and with Baker's antifol and geldanamycin. Use of RNA interference reduced ABCB1 mRNA levels and concomitantly increased sensitivity to these two drugs, as expected for ABCB1 substrates. In addition, specific silencing of ABCB5 by small interfering RNA increased sensitivity to several drugs in melanoma cells, implicating ABCB5 as a novel chemoresistance factor. Ion exchangers, ion channels, and subunits of proton and sodium pumps variably correlated with drug potency (Huang *et al.*, 2004).

Genotoxicity

The subject of genotoxicity is chromosome damage and mutations of DNA which are considered to be hallmarks of carcinogenesis. A standard genotoxicity testing system is required by regulatory agencies (Muller *et al.*, 1999). Broad mechanism based assays are not available. Currently used methods are laborious and time-consuming. For this reason experimental approaches allowing mechanism-based risk assessment are extremely valuable.

Recent progress in the development of microarrays for large-scale parallel gene expression profiling has enabled the study of genotoxic stress on a genomic scale. First studies were performed on *S. cerevisiae*. Gene expression profile analysis revealed the complexity of the genotoxic stress response (Begley and Samson, 2004). Analysis of expression profiles specific to various sources of genotoxic stress such as osmotic shock, hydrogen peroxide, amino acid starvation led to the identification of environmental stress response genes which involve 15% of the yeast genome (Gasch *et al.*, 2000, 2001; Gasch and Werner-Washburne, 2002).

DNA damaging chemicals provide agent specific expression profiles in yeast (Jelinsky *et al.*, 2000). Interestingly among the hundreds of agent specific genes, only 21 responded to all compounds in a similar fashion. It was possible to distinguish DNA-reactive and DNA-non-reactive compounds based on their gene expression profiles (Hu *et al.*, 2004). Genes involved in cell cycle regulation, DNA repair, apoptosis and cellular signaling that are distinct from other treatments appear frequently in the gene expression profiles of DNA-damaging. There is increasing evidence that gene expression analysis is capable of differentiating genotoxic mechanisms from cytotoxic mechanisms (Aubrecht and Caba, 2005; Caba *et al.*, 2005). These studies suggest that pathways, such as the p53 pathway, rather than individual genes are better predictors of genotoxic mechanisms. Clinically, these common features have been used to biomonitor human exposure to ionizing radiation (Amundson *et al.*, 2004).

Integration

Computational tools

Clustering techniques are applied to the datasets from gene expression experiments to extract toxicity signatures as phylogenetic trees. Commercially available tools include Rosetta Resolver (http://www.rosettabio.com), Gene Spring (http://www.silicongenetics.com) and Guided Analytic application (http://www.spotfire.com). Recently algorithms have been developed which relate gene expression data to protein interaction networks (Tetko *et al.*, 2005; Tornow and Mewes, 2003; Facius *et al.*, 2005). Cellular networks include metabolic or signaling pathways and pathway interactions as well as regulatory or co-expression networks. Superparamagnetic clustering has been developed as a tool to analyze collective, multi-body correlations in a genetic network (Tetko *et al.*, 2005).

Methods have been developed with the goal of visualizing global cellular mechanisms accounting for differences in gene expression. These methods relate differentially expressed genes in condition-specific, functional "signature networks" (Barabasi and Oltvai, 2004; Spirin and Mirny, 2003). These efforts have led to the development of integrated data-mining suites such as Pathway Assist (http://www.ariadnegenomics.com), PathArt (http://jubilantbiosys.com), MetaCore (http://www.genegeo.com) and Pathways Analysis (http://www.ingenuity.com).

Computational models have been developed to simulate networks that regulate transcription and metabolism in *E. coli* (Werner, 2003; Table 4.1).

Data

There is a strong need for public databases that combine gene expression profile data with associated biological, chemical and toxicological endpoints.

Table 4.1 Online resources for biological data and information for toxicogenomics study

Source	Link
TOXNET	http://toxnet.nlm.nih.gov/
NIEHS NCT (National Center for Toxicology)	http://www.niehs.nih.gov/nct/
TRC (Toxicogenomics Research Consortium)	http://www.niehs.nih.gov/nct/trc.htm
CEBS (Chemical Effects in Biological Systems)	http://www.niehs.nih.gov/nct/cebs.htm
NIEHS Microarray Group	http://dir.niehs.nih.gov/microarray/home.htm
MGED (Microarray Gene Expression Data Society)	http://www.mged.org/
EBI (European Bioinformatics Institute)	http://www.ebi.ac.uk/Information/sitemap.html
GeneCards (Weizmann Institute)	http://bioinfo.weizmann.ac.il/cards/index.html
KEGG (Kyoto Encyclopedia of Genes and Genomes)	http://www.genome.ad.jp/kegg

Although efforts to compile ADME/Tox data have been limited, some databases such as PharmaGKB (Oliver *et al.*, 2002), human membrane transporter database, the nuclear receptor database (http://receptors.ucsf.edu/NR/) and the ADME-AP (Sun *et al.*, 2002) database exist. Commercial databases for metabolic data are also on offer, Metabolism (http://www.accelrys.com) or Metabolite (http://www.mdl.com).

Databases for gene expression profiles, histopathology, clinical chemistry and morphology of animal organs after treatment with drugs have been compiled. A library of responses to known toxic agents in a particular model system can function as a reference point that can be used to interpret the responses to candidates of unknown toxicity. A reference "compendium" of profiles as a resource for toxicity prediction has been collected by the company Iconix (http://www.iconixpharm.com). Such compendia have also been built by Gene Logic (http://www.genelogic.com) and CuraGen (http://www.curagen.com). Each has gathered data from hundreds of samples produced from short-term exposures of agents at pharmacological and toxicological dose levels. Customers of both companies can access the respective databases to classify the mode-of-action of novel agents of interest. Public microarray databases (such as ArrayExpress (Europe) (Brazma *et al.*, 2003), GEO (US) (Ball *et al.*, 2004, 2005) or CIBEX (Japan) (Ikeo *et al.*, 2003)) collect data according to common guide lines (Minimum Information About a Microarray Experiment (MIAME) guidelines) (Brazma *et al.*, 2001).

To promote a systems biology approach to understanding the biological effects of environmental chemicals and stressors, the Chemical Effects in Biological Systems (CEBS) knowledgebase is being developed at the NIH to collect data from many complex data sources in a way that will allow extensive and complex queries from users. Unified data representation will occur through a systems-biology object model (a system for managing diverse -omics and toxicology/pathology data formats) that incorporates current standards for data capture and exchange (CEBS SysBio-OM) (Xirasagar *et al.*, 2004). Data streams will include gene expression, protein expression, interaction and changes in low-molecular-weight metabolite levels on agents studied, in addition to associated toxicology, histopathology and pertinent literature. Standardized procedures, protocols, data formats, and assessment methods will be used to ensure that data meet a uniform high level of quality. Relational and descriptive compendia will be included on toxicologically important genes, groups of genes, SNPs, and mutants and their functional phenotypes. CEBS will be fully searchable by compound, structure, toxicity, pathology, gene, gene group, SNP, pathway, and network. CEBS will be linked extensively to other databases and to Web genomics and proteomics resources, providing users the suite of information and tools needed to fully interpret toxicogenomics data.

Computational models are available for predicting ADME/Tox properties using software for either custom-model building and or pre-built modeling suites [Cerius$^{2™}$ ADME (http://www.accelrys.com) and KnowItAll™ (http://www.bio-rad.com)]. These programs are based on quantitative structure–activity relationships (QSARs) that use descriptors based on molecular structure and apply computational algorithms to correlate the key descriptors to the biological activity. The accumulation of drug-metabolism data from the literature has resulted in expert systems for predicting metabolism with products such as MetabolExpert™ (http://www.compudrug.com/), META™ (http://www.multicase.com/) and METEOR™ (http://www.chem.leeds.ac.uk/luk/), with the caveat that these contain data from many different mammalian species. Simulation methods have also been developed, including physiologically based pharmacokinetic modeling (PBPK) and methods such as Cloe PK™ (http://www.cyprotex.com), GastroPlus™ (http://www.simulations-plus.com), Simcyp™ (http://www.simcyp.com/) and others that include toxicokinetic methods. PBPK approaches can be used with either empirical data, *in vitro* data or in silico predictions to derive human pharmacokinetic parameters such as area under the curve (AUC) (Leahy, 2003). By contrast, computational approaches for predicting toxicity are studied infrequently but are complementary to research on ADME parameters. These methods for individual toxicology properties tend to be rule-based systems such as DEREK™ (http://www.chem.leeds.ac.uk/luk/), Hazard Expert™ (http://www.compudrug.com/), LeadScope™ (http://www.leadscope.com/) or the mechanistic methods COMPACT [65] and MultiCASE™ (http://www.multicase.com/).

References

Aebersold, R., and Mann, M. (2003). Mass spectrometry-based proteomics. Nature *422*, 198–207.

Amin, R.P., Hamadeh, H.K., Bushel, P.R., Bennett, L., Afshari, C.A., and Paules, R.S. (2002). Genomic interrogation of mechanism(s) underlying cellular responses to toxicants. Toxicology *181–182*, 555–563.

Amundson, S.A., Grace, M.B., McLeland, C.B., Epperly, M.W., Yeager, A., Zhan, Q., Greenberger, J.S., and Fornace, A.J., Jr. (2004). Human in vivo radiation-induced biomarkers: gene expression changes in radiotherapy patients. Cancer Res. *64*, 6368–6371.

Arimoto, R., Prasad, M.A., and Gifford, E.M. (2005). Development of CYP3A4 inhibition models: comparisons of machine-learning techniques and molecular descriptors. J. Biomol. Screen. *10*, 197–205.

Aubrecht, J., and Caba, E. (2005). Gene expression profile analysis: An emerging approach to investigate mechanisms of genotoxicity. Pharmacogenomics *6*, 419–428.

Ball, C.A., Awad, I. A., Demeter, J., Gollub, J., Hebert, J.M., Hernandez-Boussard, T., Jin, H., Matese, J.C., Nitzberg, M., Wymore, F., *et al.* (2005). The Stanford Microarray Database accommodates additional microarray platforms and data formats. Nucleic Acids Res. *33*, D580–582.

Ball, C.A., Brazma, A., Causton, H., Chervitz, S., Edgar, R., Hingamp, P., Matese, J. C., Parkinson, H., Quackenbush, J., Ringwald, M., *et al.* (2004). Submission of microarray data to public repositories. PLoS Biol. *2*, E317.

Barabasi, A.L., and Oltvai, Z.N. (2004). Network biology: understanding the cell's functional organization. Nat. Rev. Genet. *5*, 101–113.

Begley, T.J., and Samson, L.D. (2004). Network responses to DNA damaging agents. DNA Repair (Amst) *3*, 1123–1132.
Brazma, A., Hingamp, P., Quackenbush, J., Sherlock, G., Spellman, P., Stoeckert, C., Aach, J., Ansorge, W., Ball, C. A., Causton, H.C., *et al.* (2001). Minimum information about a microarray experiment (MIAME)-toward standards for microarray data. Nat. Genet. *29*, 365–371.
Brazma, A., Ikeo, K., and Tateno, Y. (2003). [Standardization of microarray experiment data]. Tanpakushitsu Kakusan Koso *48*, 280–285.
Bushel, P.R., Hamadeh, H.K., Bennett, L., Green, J., Ableson, A., Misener, S., Afshari, C.A., and Paules, R.S. (2002). Computational selection of distinct class- and subclass-specific gene expression signatures. J. Biomed. Inform. *35*, 160–170.
Byvatov, E., Fechner, U., Sadowski, J., and Schneider, G. (2003). Comparison of support vector machine and artificial neural network systems for drug/nondrug classification. J. Chem. Inf. Comput. Sci. *43*, 1882–1889.
Caba, E., Dickinson, D.A., Warnes, G.R., and Aubrecht, J. (2005). Differentiating mechanisms of toxicity using global gene expression analysis in Saccharomyces cerevisiae. Mutat. Res. *575*, 34–46.
de Longueville, F., Atienzar, F. A., Marcq, L., Dufrane, S., Evrard, S., Wouters, L., Leroux, F., Bertholet, V., Gerin, B., Whomsley, R., *et al.* (2003). Use of a low-density microarray for studying gene expression patterns induced by hepatotoxicants on primary cultures of rat hepatocytes. Toxicol. Sci. *75*, 378–392.
Ellinger-Ziegelbauer, H., Stuart, B., Wahle, B., Bomann, W., and Ahr, H.J. (2005). Comparison of the expression profiles induced by genotoxic and nongenotoxic carcinogens in rat liver. Mutat Res.
Facius, A., Englbrecht, C., Birzele, F., Groscurth, A., Benjamin, S., Wanka, S., and Mewes, W. (2005). PRIME: a graphical interface for integrating genomic/proteomic databases. Proteomics *5*, 76–80.
Gasch, A.P., Huang, M., Metzner, S., Botstein, D., Elledge, S. J., and Brown, P.O. (2001). Genomic expression responses to DNA-damaging agents and the regulatory role of the yeast ATR homolog Mec1p. Mol. Biol. Cell *12*, 2987–3003.
Gasch, A.P., Spellman, P.T., Kao, C. M., Carmel-Harel, O., Eisen, M.B., Storz, G., Botstein, D., and Brown, P.O. (2000). Genomic expression programs in the response of yeast cells to environmental changes. Mol. Biol. Cell *11*, 4241–4257.
Gasch, A.P., and Werner-Washburne, M. (2002). The genomics of yeast responses to environmental stress and starvation. Funct. Integr. Genomics *2*, 181–192.
Golub, T.R., Slonim, D.K., Tamayo, P., Huard, C., Gaasenbeek, M., Mesirov, J. P., Coller, H., Loh, M.L., Downing, J.R., Caligiuri, M.A., *et al.* (1999). Molecular classification of cancer: class discovery and class prediction by gene expression monitoring. Science *286*, 531–537.
Griffin, J.L., and Bollard, M.E. (2004). Metabonomics: its potential as a tool in toxicology for safety assessment and data integration. Curr. Drug Metab. *5*, 389–398.
Hamadeh, H.K., Bushel, P.R., Jayadev, S., DiSorbo, O., Bennett, L., Li, L., Tennant, R., Stoll, R., Barrett, J.C., Paules, R.S., *et al.* (2002a). Prediction of compound signature using high density gene expression profiling. Toxicol. Sci. *67*, 232–240.
Hamadeh, H.K., Bushel, P.R., Jayadev, S., Martin, K., DiSorbo, O., Sieber, S., Bennett, L., Tennant, R., Stoll, R., Barrett, J.C., *et al.* (2002b). Gene expression analysis reveals chemical-specific profiles. Toxicol. Sci. *67*, 219–231.
Hamadeh, H.K., Knight, B.L., Haugen, A.C., Sieber, S., Amin, R.P., Bushel, P.R., Stoll, R., Blanchard, K., Jayadev, S., Tennant, R. W., *et al.* (2002c). Methapyrilene toxicity: anchorage of pathologic observations to gene expression alterations. Toxicol. Pathol. *30*, 470–482.

Hastie, T., Tibshirani, R., Botstein, D., and Brown, P. (2001). Supervised harvesting of expression trees. Genome Biol. 2, RESEARCH0003.

Heijne, W.H., Kienhuis, A.S., van Ommen, B., Stierum, R.H., and Groten, J.P. (2005). Systems toxicology: applications of toxicogenomics, transcriptomics, proteomics and metabolomics in toxicology. Expert Rev. Proteomics *2*, 767–780.

Heinloth, A.N., Irwin, R.D., Boorman, G.A., Nettesheim, P., Fannin, R.D., Sieber, S.O., Snell, M.L., Tucker, C.J., Li, L., Travlos, G.S., *et al.* (2004). Gene expression profiling of rat livers reveals indicators of potential adverse effects. Toxicol. Sci. *80*, 193–202.

Hu, T., Gibson, D.P., Carr, G.J., Torontali, S.M., Tiesman, J.P., Chaney, J.G., and Aardema, M.J. (2004). Identification of a gene expression profile that discriminates indirect-acting genotoxins from direct-acting genotoxins. Mutat. Res. *549*, 5–27.

Huang, Y., Anderle, P., Bussey, K.J., Barbacioru, C., Shankavaram, U., Dai, Z., Reinhold, W.C., Papp, A., Weinstein, J.N., and Sadee, W. (2004). Membrane transporters and channels: role of the transportome in cancer chemosensitivity and chemoresistance. Cancer Res. *64*, 4294–4301.

Ikeo, K., Ishi-i, J., Tamura, T., Gojobori, T., and Tateno, Y. (2003). CIBEX: center for information biology gene expression database. C. R. Biol. *326*, 1079–1082.

Jelinsky, S.A., Estep, P., Church, G.M., and Samson, L.D. (2000). Regulatory networks revealed by transcriptional profiling of damaged *Saccharomyces cerevisiae* cells: Rpn4 links base excision repair with proteasomes. Mol. Cell Biol. *20*, 8157–8167.

Kaput, J., and Rodriguez, R.L. (2004). Nutritional genomics: the next frontier in the postgenomic era. Physiol. Genomics *16*, 166–177.

Keun, H.C. (2005). Metabonomic modeling of drug toxicity. Pharmacol. Ther.

Khan, J., Wei, J.S., Ringner, M., Saal, L.H., Ladanyi, M., Westermann, F., Berthold, F., Schwab, M., Antonescu, C. R., Peterson, C., and Meltzer, P.S. (2001). Classification and diagnostic prediction of cancers using gene expression profiling and artificial neural networks. Nat. Med. *7*, 673–679.

Kier, L.D., Neft, R., Tang, L., Suizu, R., Cook, T., Onsurez, K., Tiegler, K., Sakai, Y., Ortiz, M., Nolan, T., *et al.* (2004). Applications of microarrays with toxicologically relevant genes (tox genes) for the evaluation of chemical toxicants in Sprague Dawley rats in vivo and human hepatocytes in vitro. Mutat. Res. *549*, 101–113.

Kramer, J.A., Curtiss, S.W., Kolaja, K.L., Alden, C.L., Blomme, E.A., Curtiss, W.C., Davila, J.C., Jackson, C.J., and Bunch, R.T. (2004). Acute molecular markers of rodent hepatic carcinogenesis identified by transcription profiling. Chem. Res. Toxicol. *17*, 463–470.

Krishnan, V.G., and Westhead, D.R. (2003). A comparative study of machine-learning methods to predict the effects of single nucleotide polymorphisms on protein function. Bioinformatics *19*, 2199–2209.

Lahm, H.W., and Langen, H. (2000). Mass spectrometry: a tool for the identification of proteins separated by gels. Electrophoresis *21*, 2105–2114.

Leahy, D.E. (2003). Progress in simulation modelling for pharmacokinetics. Curr. Top. Med. Chem. *3*, 1257–1268.

Leighton, J.K. (2005). Application of emerging technologies in toxicology and safety assessment: regulatory perspectives. Int. J. Toxicol. *24*, 153–155.

Lio, P. (2003). Wavelets in bioinformatics and computational biology: state of art and perspectives. Bioinformatics *19*, 2–9.

Lord, P.G. (2004). Progress in applying genomics in drug development. Toxicol. Lett. *149*, 371–375.

Lord, P.G., and Papoian, T. (2004). Genomics and drug toxicity. Science *306*, 575.

Merrick, B.A., and Madenspacher, J.H. (2005). Complementary gene and protein expression studies and integrative approaches in toxicogenomics. Toxicol. Appl. Pharmacol. *207*, 189–194.

Morgan, K.T. (2002). Gene expression analysis reveals chemical-specific profiles. Toxicol. Sci. *67*, 155–156.

Muller, L., Kikuchi, Y., Probst, G., Schechtman, L., Shimada, H., Sofuni, T., and Tweats, D. (1999). ICH-harmonised guidances on genotoxicity testing of pharmaceuticals: evolution, reasoning and impact. Mutat. Res. *436*, 195–225.

Oliver, D.E., Rubin, D. L., Stuart, J.M., Hewett, M., Klein, T. E., and Altman, R.B. (2002). Ontology development for a pharmacogenetics knowledge base. Pac. Symp. Biocomput. 65–76.

Pognan, F. (2004). Genomics, proteomics and metabonomics in toxicology: hopefully not 'fashionomics'. Pharmacogenomics 5, 879–893.

Richburg, J.H., Johnson, K.J., Schoenfeld, H.A., Meistrich, M.L., and Dix, D.J. (2002). Defining the cellular and molecular mechanisms of toxicant action in the testis. Toxicol. Lett. *135*, 167–183.

Robertson, D.G. (2005). Metabonomics in toxicology: a review. Toxicol. Sci. *85*, 809–822.

Scholkopf, B., Smola, A. J., Williamson, R. C., and Bartlett, P. L. (2000). New support vector algorithms. Neural Comput. *12*, 1207–1245.

Smola, A.J., Scholkopf, B., and Muller, K. R. (1998). The connection between regularization operators and support vector kernels. Neural Netw. *11*, 637–649.

Spirin, V., and Mirny, L. A. (2003). Protein complexes and functional modules in molecular networks. Proc. Natl. Acad. Sci. USA *100*, 12123–12128.

Steiner, G., Suter, L., Boess, F., Gasser, R., de Vera, M. C., Albertini, S., and Ruepp, S. (2004). Discriminating different classes of toxicants by transcript profiling. Environ. Health Perspect. *112*, 1236–1248.

Sun, L. Z., Ji, Z. L., Chen, X., Wang, J.F., and Chen, Y.Z. (2002). ADME-AP: a database of ADME associated proteins. Bioinformatics *18*, 1699–1700.

Tennant, R. W. (2002). The National Center for Toxicogenomics: using new technologies to inform mechanistic toxicology. Environ. Health Perspect. *110*, A8–10.

Tetko, I. V., Facius, A., Ruepp, A., and Mewes, H.W. (2005). Super paramagnetic clustering of protein sequences. BMC Bioinformatics 6, 82.

Tornow, S., and Mewes, H. W. (2003). Functional modules by relating protein interaction networks and gene expression. Nucleic Acids Res. *31*, 6283–6289.

Vapnik, V., and Chapelle, O. (2000). Bounds on error expectation for support vector machines. Neural Comput. *12*, 2013–2036.

Wang, Y., Tetko, I.V., Hall, M.A., Frank, E., Facius, A., Mayer, K.F., and Mewes, H.W. (2005). Gene selection from microarray data for cancer classification—a machine learning approach. Comput. Biol. Chem. *29*, 37–46.

Waring, J.F., Cavet, G., Jolly, R.A., McDowell, J., Dai, H., Ciurlionis, R., Zhang, C., Stoughton, R., Lum, P., Ferguson, A., *et al.* (2003). Development of a DNA microarray for toxicology based on hepatotoxin-regulated sequences. EHP Toxicogenomics *111*, 53–60.

Waring, J.F., Gum, R., Morfitt, D., Jolly, R.A., Ciurlionis, R., Heindel, M., Gallenberg, L., Buratto, B., and Ulrich, R.G. (2002). Identifying toxic mechanisms using DNA microarrays: evidence that an experimental inhibitor of cell adhesion molecule expression signals through the aryl hydrocarbon nuclear receptor. Toxicology *181–182*, 537–550.

Waters, M., Boorman, G., Bushel, P., Cunningham, M., Irwin, R., Merrick, A., Olden, K., Paules, R., Selkirk, J., Stasiewicz, S., *et al.* (2003). Systems toxicology and the Chemical Effects in Biological Systems (CEBS) knowledge base. EHP Toxicogenomics *111*, 15–28.

Wei, L., Yang, Y., Nishikawa, R.M., and Jiang, Y. (2005). A study on several machine-learning methods for classification of malignant and benign clustered microcalcifications. IEEE Trans. Med. Imaging *24*, 371–380.

Werner, E. (2003). In silico multicellular systems biology and minimal genomes. Drug Discov. Today *8*, 1121–1127.

Witkamp, R.F. (2005). Genomics and systems biology—how relevant are the developments to veterinary pharmacology, toxicology and therapeutics? J. Vet. Pharmacol. Ther. *28*, 235–245.

Xirasagar, S., Gustafson, S., Merrick, B.A., Tomer, K.B., Stasiewicz, S., Chan, D.D., Yost, K.J., 3rd, Yates, J. R., 3rd, Sumner, S., Xiao, N., and Waters, M. D. (2004). CEBS object model for systems biology data, SysBio-OM. Bioinformatics *20*, 2004–2015.

Microarray Gene Expression Atlases

5

John C. Castle, Chris J. Roberts, Chun Cheng, and Jason M. Johnson

Abstract

Gene expression atlases are systematically compiled RNA expression measurements intended to show the relative abundance of gene transcripts across diverse collections of samples. Generation of a gene expression atlas is often a large, expensive project. However, once compiled, these atlases are excellent references and empower molecular biologists, bioinformaticians, and statisticians to examine tissue and gene expression to improve our understanding of the molecular processes driving cell function. They can be used to answer many questions, from simple to complex: examples of biological questions include straightforward queries such as where genes are expressed and what genes are expressed in a sample and extend to more complex queries involving gene–gene and tissue–tissue expression correlations and gene functional/pathway assignments. Several public gene expression atlases exist and we present examples from our research.

Introduction

Gene expression "atlases" are compilations of systematically collected RNA expression measurements showing relative abundance of gene transcripts across diverse collections of samples and serve as references for many different types of biological questions. Atlas samples are often derived from distinct organs from throughout the body and can be made from human tissues; tissues from other species, such as mouse, rat, and dog; or from collections of cell lines and tumors. Atlases can be generated using many assays: microarrays in particular are well suited for atlas generation as microarray profiling is high throughput, both in the number of genes simultaneously monitored and the ability to profile multiple samples. Many gene expression atlases are currently available in the public domain. Table 5.1 is a partial collection of large, publicly available microarray datasets. Here, we present examples drawn from our work of the types of research that can be done using gene expression atlases.

Table 5.1 Publicly available microarray atlases

Species	Name	Genes	Tissues	Link to data
Human	GNF Human Atlas (Su *et al.*, 2004)	22,000	79	http://www.ncbi.nlm.nih.gov/geo/gds/gds_browse.cgi?gds = 596
	NCI (Son *et al.*, 2005)	19,000	160	http://www.genome.org/cgi/content/full/15/3/443/DC1
	Rosetta PTA (Schadt *et al.*, 2004)	20,000	60	http://www.ncbi.nlm.nih.gov/geo/query/acc.cgi?acc = GSE918
	Rosetta Jchip (Johnson *et al.*, 2003)	10,000	52	http://www.ncbi.nlm.nih.gov/geo/query/acc.cgi?acc = GSE740
Mouse	GNF Mouse Atlas (Su *et al.*, 2004)	12,500	45	http://www.ncbi.nlm.nih.gov/geo/gds/gds_browse.cgi?gds = 182
	Toronto Mouse Atlas (Zhang *et al.*, 2004)	40,000	55	http://hugheslab.med.utoronto.ca/Zhang/
Rat	GNF Rat Atlas (Walker *et al.*, 2004)	8,800	Four strains, 30 neurological tissues	http://www.ncbi.nlm.nih.gov/geo/gds/gds_browse.cgi?gds = 589
Cell line	NCI60 (Ross *et al.*, 2000)	8,000	60 cell lines	http://genome-www.stanford.edu/nci60/

Primary uses

Where is my gene expressed?

The single most common use of atlases is to determine where a gene is expressed, such as where in an organism or within a set of cell lines or tumor types. A gene can come to the attention of a researcher for a variety of reasons, ranging from a paper in the literature to a hit in a high-throughput RNAi screen, and in many cases there is no *a priori* knowledge of where the gene is expressed. Furthermore, if there are published gene expression data in normal tissues, the profiling is often constrained to a limited number of tissues or a less-relevant species.

Atlases are also useful for examining the expression of families of genes, such as the voltage-gated sodium channels. Voltage-gated sodium channels functions to transmit signals, including transmitting pain signals from the periphery through the dorsal root ganglion (DRG) to the spinal cord. A potential intervention point for pain treatment would be to prevent this pain transmission, possibly through interruption of sodium channel signaling in the DRG. However, inhibition of sodium channel signaling in other tissues, such as muscle and heart, would need to be avoided to prevent non-desirous effects.

Thus, voltage-gated sodium channels expressed in DRG and not in other tissues are candidate targets for pain treatment. Fig. 5.1 shows expression of eight voltage-gated sodium channels across 50 cynomolgus monkey tissues, including brain and dorsal root ganglion (Raymond *et al.*, 2004). The use of non-human primate tissue allows collection of consistently dissected, disease-free, drug-free, middle-aged tissues. Heart expresses primarily *SCN5A* and skeletal muscle expresses primarily *SCN4A*; thus neither gene is a likely pain target. *SCN2A* is expressed in brain subsections, *SCN7A* is expressed in many tissue types, and *SCN1A* and *SCN8A* are expressed in both central and peripheral nervous tissues. However, *SCN9A*, *SNC10A*, and *SCN11A* are all highly expressed in monkey dorsal root ganglion. Thus, these three genes are candidate targets for pain treatment.

What genes are expressed in my tissue?

Gene expression atlases also show what genes are expressed in a specific sample, such as a cell line or brain subsection. Thus, rather than ask gene-centric questions, such as "Where is my gene expressed?" the researcher may ask tissue-centric questions, such as "What genes are expressed in my tissue?"

Insulin-making cells are found in the islets of Langerhans (pancreatic islets) and genes expressed specifically in islets may be involved in insulin generation and thus possibly involved in diabetes (McEntyre, 2004). Using our atlas of gene expression in mouse tissues, we identified a gene expressed only in mouse pancreatic islets (Fig. 5.2). This gene, a receptor, is currently in assay development for high-throughput screening for treatment of metabolic syndrome. Thus, the availability of the atlas allowed us to find a gene expressed specifically in islets.

What groups of gene are specific to my tissue?

Similarly, sets of genes can be identified as specific to a given tissue or sample. Given an atlas compiling measurements of thousands of genes in tens or hundreds of tissues, researchers can develop statistical methods for determining sets of genes enriched in a given tissue type. Two metrics include "fold-change to second highest" and "specificity." For each gene, the "fold-change to second highest" is calculated as the ratio of the intensity observed in the tissue of interest to the intensity of the highest-expressing tissue, ignoring the tissue of interest. If the fold-change is greater than one, the gene is expressed at highest levels in the tissue of interest and higher fold-change values show higher specificity. Another metric, called "specificity," can be calculated for each gene as the intensity in the tissue of interest divided by the sum of intensities for all tissues. A value of 0 implies the gene is not expressed in the tissue whereas a value of 1 implies that the gene is only expressed in the tissue.

Tissue types may be as simple as a single tissue in the atlas, such as liver, or may be more complicated, such as "brain" or "gut," in which case expression values from many tissues are first combined before comparison to other tissues.

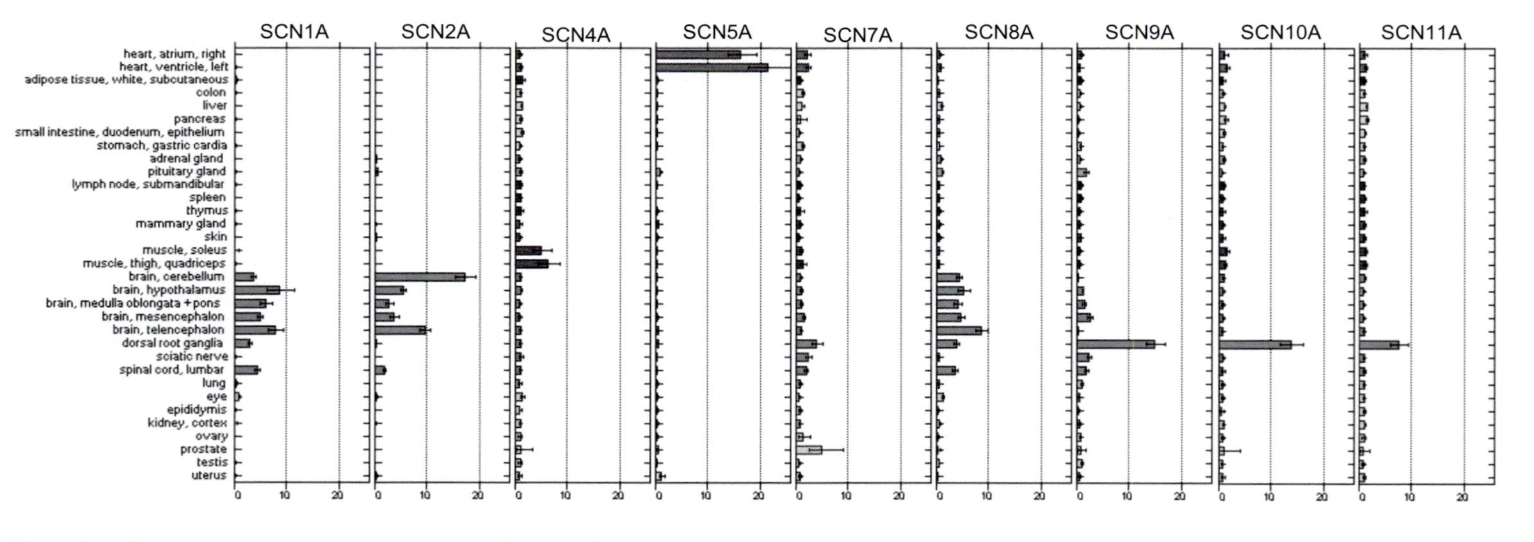

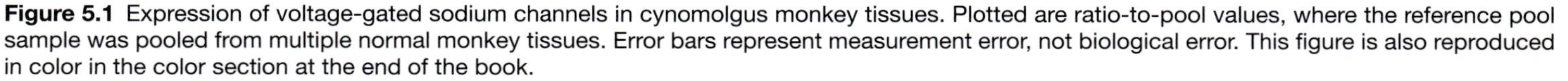

Figure 5.1 Expression of voltage-gated sodium channels in cynomolgus monkey tissues. Plotted are ratio-to-pool values, where the reference pool sample was pooled from multiple normal monkey tissues. Error bars represent measurement error, not biological error. This figure is also reproduced in color in the color section at the end of the book.

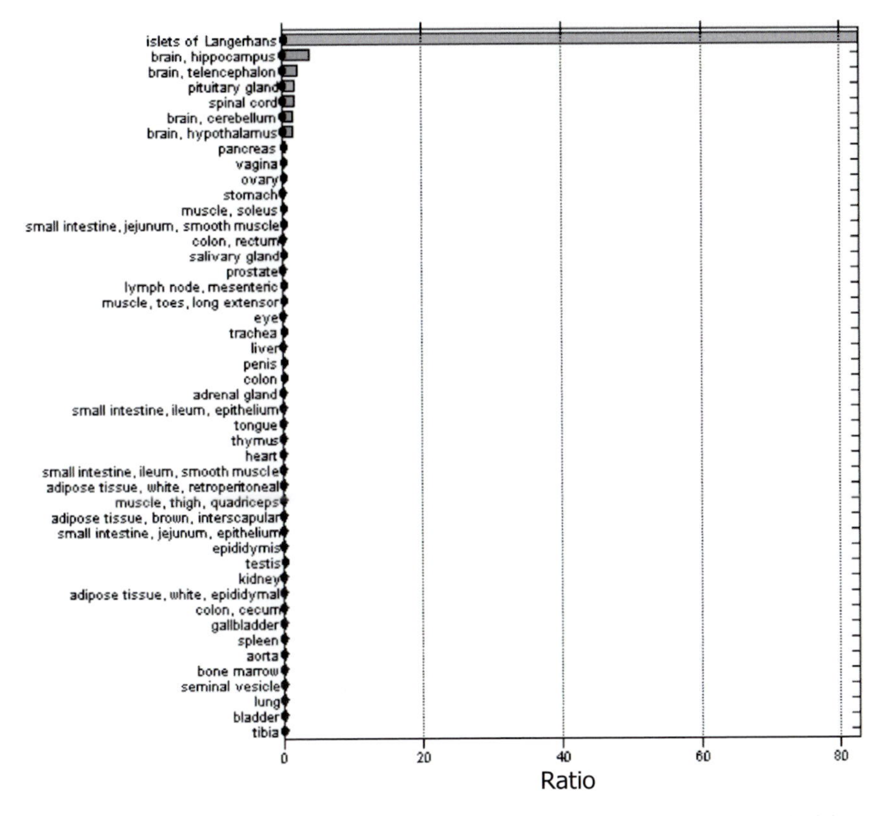

Figure 5.2 Expression of a gene expressed specifically in mouse pancreatic islets. Plotted are ratio-to-pool values, where the reference pool sample was pooled from multiple normal mouse tissues.

Similarly, genes can be enriched in a tissue or tissue combination relative to either all of the tissues in the atlas or relative to only a subset of tissues. For example, an atlas could be examined to find cerebellum-specific genes not found elsewhere in the body, excluding other brain tissues, or to find cerebellum-specific genes not found elsewhere in the body, including other brain tissues.

Fig. 5.3 shows 574 genes that show enriched expression in monkey liver. Some genes show expression to a lesser degree in other tissues, such as gallbladder, but all show highest expression in liver. These include known liver-specific genes, such as cytochrome P450s, and previously uncharacterized genes. Specific sets like these can be used for many purposes, from target identification and prioritization (e.g., islet-specific genes) to understanding transcriptional activity in the liver (e.g., molecular pathways) to searching subsequent experiments for evidence of liver contamination (e.g., sample quality control). For instance, if an adipose sample, in an unrelated experiment, showed expression of many muscle-specific genes, it would suggest possible muscle contamination during dissection.

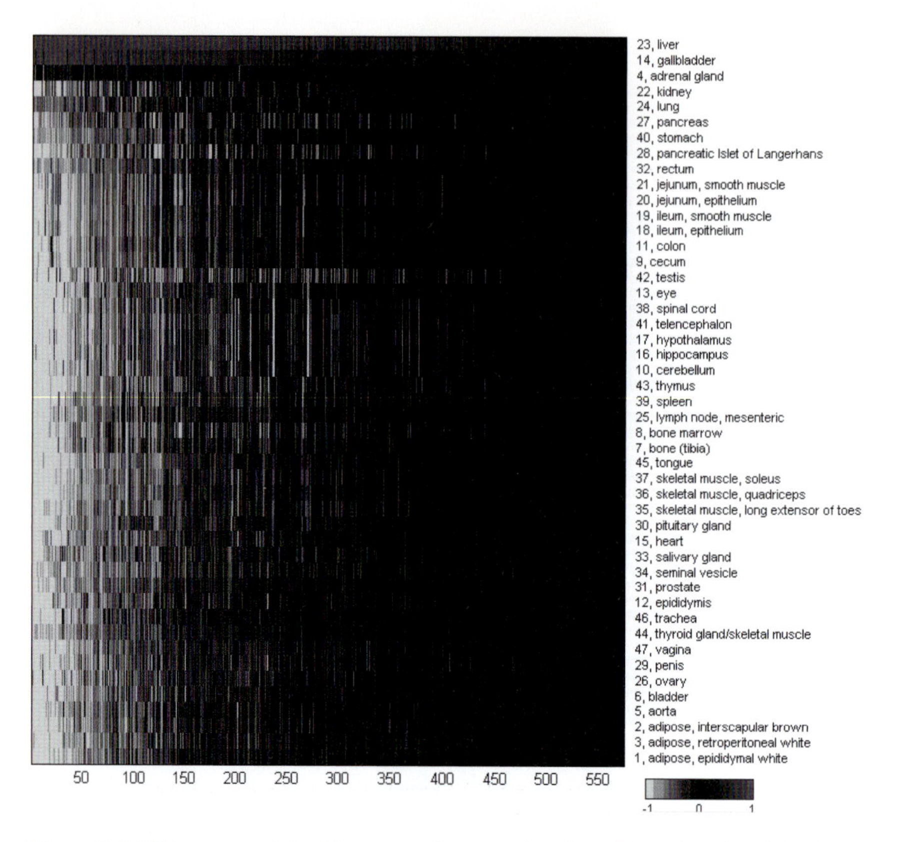

Figure 5.3 574 genes enriched in cynomolgus monkey liver. Plotted are log 10 ratio-to-pool values, where the reference pool sample was pooled from multiple normal monkey tissues. Values range from –2 to 2 and are clipped at –1 to 1 for this figure. Genes are ordered based on specificity and tissues are clustered using agglomerative hierarchical clustering. This figure is also reproduced in color in the color section at the end of the book.

Biomarkers: which genes are expressed specifically in condition X?

Just as one can examine an atlas for genes expressed in a certain tissue, or tissues expressing a certain gene, one can also query an atlas for genes expressed in a specific condition but not in other conditions. For example, if an atlas contains both disease and disease-free samples, this method can be used to search for disease biomarkers.

Bone mineral density (BMD) is the standard biomarker for bone growth or loss; however, changes in BMD occur over a long period of time (i.e., weeks to months). An acute biomarker to track bone loss and formation and monitor deposition rates would be of great benefit to clinical trials for osteoporosis, especially if it could be monitored in the periphery (e.g., a secreted gene). Furthermore, different bones (e.g., vertebra and hip) are composed of different proportions of bone types (e.g., trabecular and cortical bone). To aid in the identification of genes that might serve as bone growth biomarkers in specific bone subsections, we included both tibia diaphysis (the long, central portion of the tibia, composed of cortical bone) and tibia metaphysis (proximal end of the tibia, composed of trabecular bone) in the monkey atlas. To find biomarkers, we first identified genes whose expression was specific to either of these tissues and that encoded secreted proteins: five genes meet these criteria (Fig. 5.4) and are thus candidate bone growth markers. Furthermore, two of these bone-specific, secreted genes are expressed in diaphysis tissue and not in metaphysis tissue. Such sub-bone type specific genes may be useful for monitoring the effects of compounds with growth effects targeted to specific bone types (e.g., vertebra and hip) and all five genes are currently in development for use in clinical settings. Thus, we were able to analyze atlas data to find candidate clinical biomarkers.

Additional uses

The examples previously described can be described as extensions of two simple but powerful queries: what tissues express a particular gene and what genes are expressed in a particular tissue. More complicated analyses are also possible and we highlight a few here.

Tissue correlation comparisons

Large gene expression atlases allow correlations of tissues against tissues, determining the similarity of tissues, and genes against genes, determining the similarity of genes. One application of this calculation is as a quality control method to test tissue annotation. For example, when we generated our first atlas, we clustered all tissues based on their tissue-tissue correlations. Tissue clustering showed ileum, a section of the small intestine, clustering with brain tissues and retina clustering with duodenum and jejunum, two other small intestine sections. After further examination of hybridization logs, we were able to confirm that the retina and ileum samples had been switched. Similarly, these types of analyses can be used to determine the purity of individual tissues. After profiling a bone sample, we examined clustering between this sample and samples already included in the atlas. The comparison indicated possible contamination of the purchased bone sample with bone marrow and muscle, suggesting replacement of the sample with a new sample. A final application of this analysis is determining the similarity of gene expression level between tissues. Cell lines, for example, cluster with each other and not with their parental tissues, demonstrating the large molecular change they undergo during immortalization.

Gene correlation comparisons

Comparisons between gene expression profiles aides in gene function and pathway determination. For example, voltage-gated sodium channels expressed in dorsal-root-ganglia are more likely to be involved in pain transmission than heart-expressed voltage-gated sodium channels. So, given a gene with well-

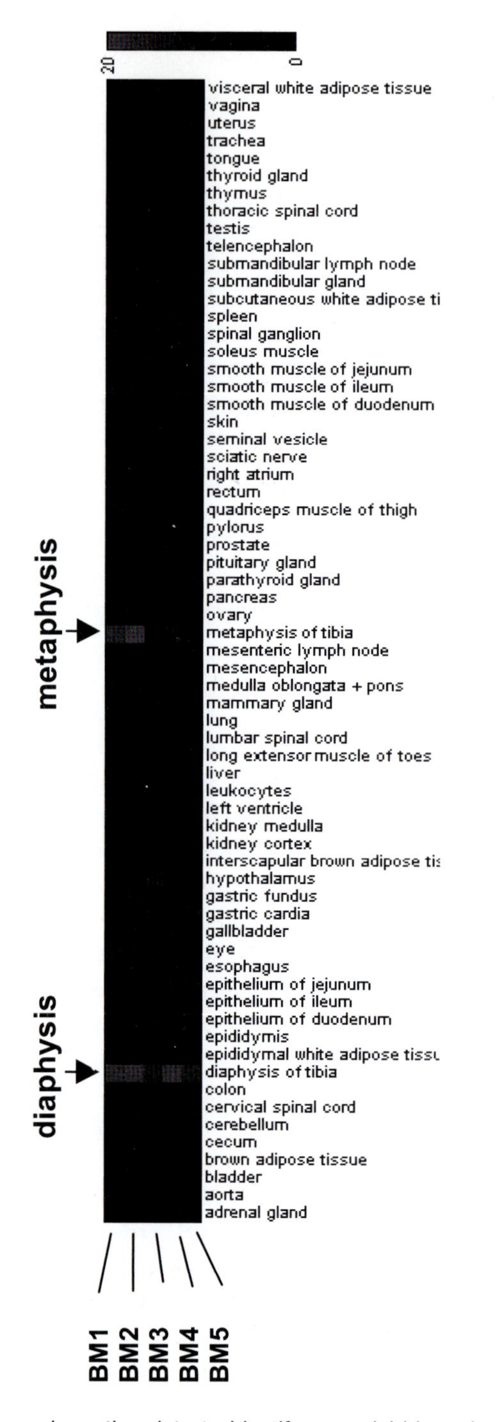

Figure 5.4 Cynomolgus monkey atlas data to identify potential biomarkers for bone growth. Plotted are log 10 ratio-to-pool values, where the reference pool sample was pooled from multiple normal monkey tissues. This figure is also reproduced in color in the color section at the end of the book.

defined function, an expression atlas allows users to identify highly correlating genes that are similarly expressed and thus are likely involved in similar molecular pathways.

Species correlation comparisons

The use of non-human model species is essential to drug discovery; however, species differences can lead to non-efficacious treatments (e.g., drugs that treat depression in mice but not humans). Gene expression atlases compiled from different species can be used to compare species, both comparing individual genes across species, determining whether homologous genes are similarly expressed, and comparing individual tissues across species, determining if similar tissues express similar genes and have similar function between species.

Fig. 5.5A and B show correlations between and clustering of human, monkey, and rat tissues based on orthologous genes. Most similar tissues show similar expression profiles and thus co-cluster. If species differences played a larger role in determining gene expression, we might expect some tissues to be more similar within a species, such as rat heart and rat skeletal muscle, than across species, such as monkey with human. However, examining the clustering shows that rat, human, and monkey skeletal muscle tissues group together, and rat, human, and monkey heart tissues group together, and only at a lower similarity do these two groups cluster together. This clustering demonstrates that in general tissue organ differences (e.g., skeletal muscle versus heart) are more significant than species differences (e.g., rat versus monkey).

While most tissues cluster with their cross-species partners, exceptions exist. Brain subsections cluster within species before clustering with their cross-species tissue, possibly as a result of true species molecular differences. Thus, comparing atlases from different species reveals molecular similarities and differences between species and identification of genes with similar expression patterns across species adds confidence to extrapolating drug discovery results from one species to a second.

Transcription factors and targets

Genes with similar expression patterns often have similar function, and these sets of genes can be examined in an atlas. For example, genes involved in certain gene ontology classes, such as sarcomere, myofibril, and "regulation of muscle contraction," are upregulated in human muscle tissues, including skeletal muscle and heart. Similarly, the targets of transcript factors should have similar expression patterns. Fig. 5.6 shows the mRNA expression of transcription factor E2F1 versus the expression of several mRNA E2F1 targets in an atlas of 50 human tissues. E2F1 target genes are determined from literature. When E2F1 is expressed at relatively low levels, the targets show considerable variation in expression, presumably because each gene also responds to multiple transcription factors. However, when E2F1 is expressed at higher levels, the E2F1 targets are

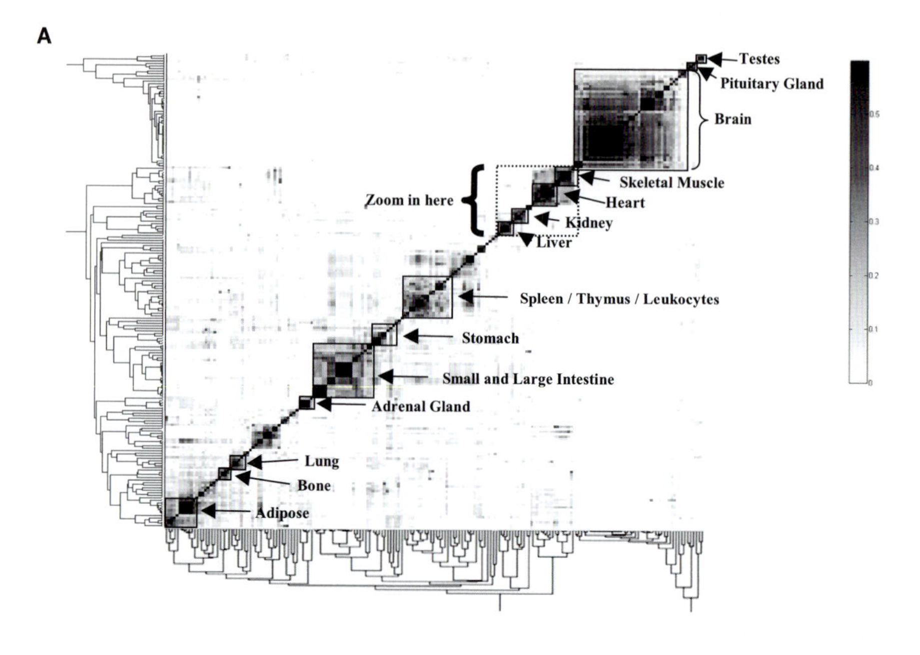

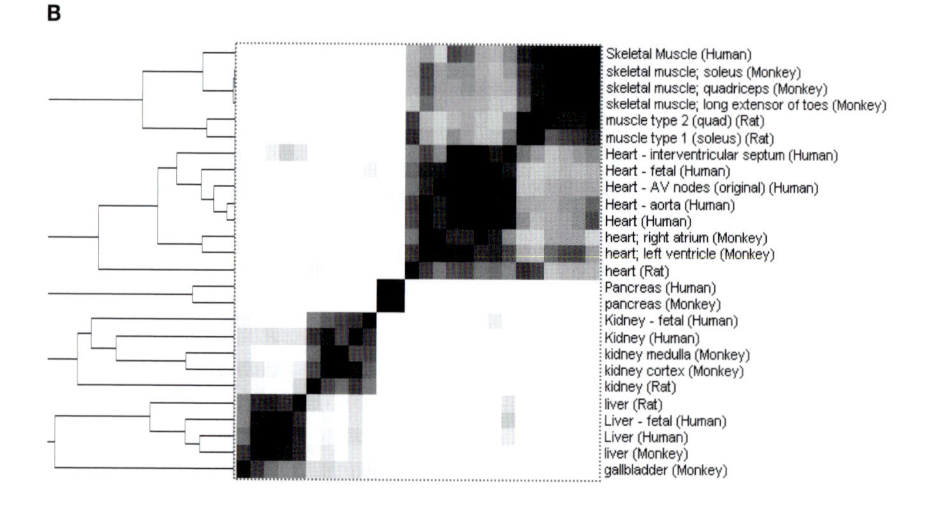

Figure 5.5 Co-clustering of human, monkey, and rat tissue profiles shows conservation of tissue-specific genes across species. 86 human, 65 monkey, and 38 rat tissue profiles are clustered using hierarchical agglomerative clustering on both axes based on the expression of 12,000 orthologous genes. Values used are ratio-to-pool, where the pool is a species-specific mixture of normal tissues. Color represents correlation, with correlation capped at 0.6. (A) All tissues examined; (B) a zoom view into a smaller region, as indicated by the dashed box. This figure is also reproduced in color in the color section at the end of the book.

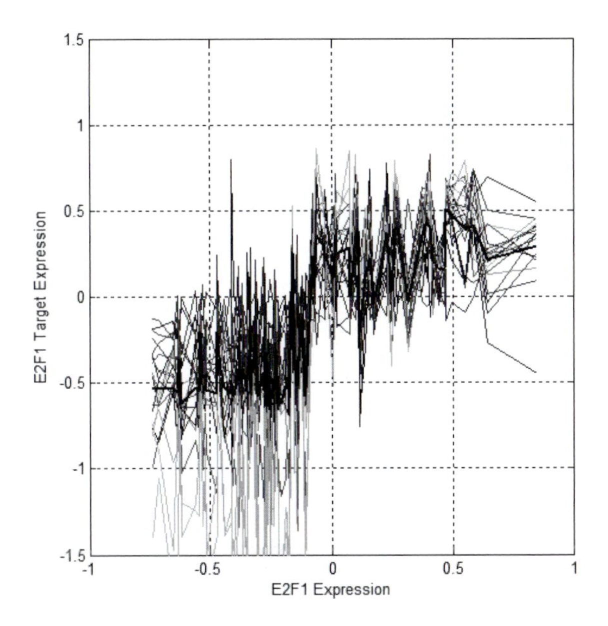

Figure 5.6 The expression of 21 targets of transcription factor E2F1 (y-axis) are plotted against the expression of transcription factor E2F1 (x-axis). The 21 E2F1 targets are determined from literature. Plotted values are log 10 ratio-to-reference pool values determined from 105 human tissues and cell lines. The reference pool is a mixture of normal human tissues. Each line represents the values for a single gene. The thick line shows the average value, across the 21 genes in each of the 105 tissues. This figure is also reproduced in color in the color section at the end of the book.

also expressed at higher levels. These studies could be extended by combining gene co-expression analysis with transcription factor binding site predictions, especially with incorporation of cross-species expression and sequence conservation, to create a powerful method for molecular pathway interrogation.

Conclusion

Creation of a gene expression atlas is a large, expensive project. However, once compiled, these atlases can be used to answer many questions, from simple to complex, and are excellent references. They enable molecular biologists, bioinformaticians, and statisticians to examine tissue and gene expression to improve our understanding of the molecular processes driving cell function.

References

Johnson, J.M., Castle, J., Garrett-Engele, P., Kan, Z., Loerch, P.M., Armour, C.D., Santos, R., Schadt, E.E., Stoughton, R., and Shoemaker, D.D. (2003). Genome-wide survey of human alternative pre-mRNA splicing with exon junction microarrays. Science *302*, 2141–2144.

McEntyre, L.D. a. J. (2004). The Genetic Landscape of Diabetes. Online at http://www.ncbi.nlm.nih.gov/books/bv.fcgi?call=bv.View..ShowTOC&rid=diabetes.TOC&depth=1

Raymond, C.K., Castle, J., Garrett-Engele, P., Armour, C. D., Kan, Z., Tsinoremas, N., and Johnson, J.M. (2004). Expression of alternatively spliced sodium channel alpha-subunit genes. Unique splicing patterns are observed in dorsal root ganglia. J. Biol. Chem. *279*, 46234–46241.

Ross, D.T., Scherf, U., Eisen, M.B., Perou, C.M., Rees, C., Spellman, P., Iyer, V., Jeffrey, S. S., Van de Rijn, M., Waltham, M., *et al.* (2000). Systematic variation in gene expression patterns in human cancer cell lines. Nat. Genet. *24*, 227–235.

Schadt, E.E., Edwards, S. W., GuhaThakurta, D., Holder, D., Ying, L., Svetnik, V., Leonardson, A., Hart, K. W., Russell, A., Li, G., *et al.* (2004). A comprehensive transcript index of the human genome generated using microarrays and computational approaches. Genome Biol. *5*, R73.

Son, C.G., Bilke, S., Davis, S., Greer, B.T., Wei, J.S., Whiteford, C.C., Chen, Q.R., Cenacchi, N., and Khan, J. (2005). Database of mRNA gene expression profiles of multiple human organs. Genome Res. *15*, 443–450.

Su, A.I., Wiltshire, T., Batalov, S., Lapp, H., Ching, K. A., Block, D., Zhang, J., Soden, R., Hayakawa, M., Kreiman, G., *et al.* (2004). A gene atlas of the mouse and human protein-encoding transcriptomes. Proc. Natl. Acad. Sci. USA *101*, 6062–6067.

Walker, J.R., Su, A.I., Self, D.W., Hogenesch, J. B., Lapp, H., Maier, R., Hoyer, D., and Bilbe, G. (2004). Applications of a rat multiple tissue gene expression dataset. Genome Res. *14*, 742–749.

Zhang, W., Morris, Q.D., Chang, R., Shai, O., Bakowski, M.A., Mitsakakis, N., Mohammad, N., Robinson, M.D., Zirngibl, R., Somogyi, E., *et al.* (2004). The functional landscape of mouse gene expression. J. Biol. 3, 21.

Supervised Classification of Genes and Biological Samples

6

Adrian Tkacz, Leszek Rychlewski, Paolo Uva, and Dariusz Plewczynski

Abstract

Microarray experiments generate large volumes of gene expression data and are currently applied to elucidate a large spectrum of biological problems, in various research contexts. Although the experimental technology has undergone important progress during the last years and is near to consolidation stage, the statistical analysis and the extraction of all the potential information residing in the data, still represents a great challenge for the scientific community. Despite the availability of advanced computational tools, the choice of analytical techniques made by the data analyst has a great impact on the practical interpretation of results. A basic understanding of these computational methods is therefore needed for experimental design and meaningful data analysis. Here we present a short introduction to supervised computational methods and data analysis tools and we illustrate how these techniques can be used with practical examples taken from the recent literature.

Introduction

The completion of the sequencing of a large number of genomes presents several challenges and opportunities, with the focus on the functional classification of predicted genes. The high volume of genome-wide microarray data available to the scientific community has already significantly contributed to the elucidation of the functional role of genes contained in entire genomes.

The utility of microarrays has been demonstrated in many biomedical research areas and the importance of proper analysis is not disputable. In spite of the rapid evolution of the experimental microarray technology, the statistical analysis of the data represents a great challenge, as also demonstrated by the still high level of heterogeneity of the analytical approaches described in the various scientific works.

This chapter is focused on the illustration of various classification approaches which can be applied for analysing and interpreting of the overwhelming amount of data generated by microarray experiments.

The majority of computational methods that have evolved for microarray data analysis can all be grouped into three main approaches: "class discovery," "class comparison," and "class prediction" (Golub *et al.*, 1999).

In class discovery the objects of study are not grouped into predefined classes: this approach is therefore "unsupervised." A clustering algorithm is applied and genes and/or samples are grouped on the basis of similarity across conditions. The method starts with the definition of a distance metric to represent the distance between the genes and the samples in terms of gene expression. The algorithm then groups the objects so that the distance between objects of the same group is minimized while the distance between objects of different groups is maximized. In the case of cancer studies, this approach has been used to identify discrete subsets of disease samples on the basis of gene expression profiles. Examples of class discovery studies can be found in the works of Bittner *et al.* (2000) and Alizadeh *et al.* (2000) in which gene expression profiles of samples preclassified as advanced melanomas and B-cell lymphoma were analyzed, and sample subsets sharing specific gene expression patterns could be identified. These methods are illustrated in the first chapter of this book.

In class comparison studies the classes are defined on the basis of a prior knowledge, independently from the expression profiles. The method is therefore "supervised" and is used to determine if the expression profiles are different between the predefined classes. The typical output of this approach is a list of genes differentially expressed between the classes. Commonly used methods to assess the statistical significance of differential expression are t-test and ANOVA. The comparison of gene expression profiles of breast cancer patients who are long-term survivors with the gene expression profiles of those who have recurrent disease is an example of this type of study (Hedenfalk *et al.*, 2001).

The class prediction approach is also "supervised," as it starts from prior knowledge to group the objects of study (e.g., the biological samples) into different classes. Based on this classification, a predictor based on the gene expression is identified. The predictor is therefore applied to classify new samples based on their expression profile. These methods have great potentialities in clinical research and applications have been described for risk assessment, diagnostic testing, prognostic stratification and treatment selection (Golub *et al.*, 1999). Class prediction studies normally consist of setting up a model so that given a collection of "predictor variables" (e.g., the gene expression values of tumoral and normal biological samples) the quality of future observations can be inferred, named "response variables" (e.g., the type of a biological samples, "tumor" or "normal"). In addition these studies provide useful information for the elucidation of the relationships between response and predictor variables of the underlying prediction model.

The supervised framework can be applied to the classification of both samples (e.g., the type of disease) and genes. In the first case samples are compared to each other in order to predict the correct assignments. This approach offers the possibility of an objective and highly accurate classification of cancers that could provide clinicians with the information needed for the most appropriate treatment. The work of Golub *et al.* (1999) represents an interesting application of this technique as they obtained promising results in discovering new cancer subtypes.

The second type of supervised analysis is focused on gene classification, based on the assumption that genes of similar function share similar expression profiles across samples (see for example the work of Eisen *et al.*, 1998).

Different classification algorithms are based on data models that are often profoundly different from each other. Therefore a key criterion to favor a class prediction algorithm in comparison to others is evaluating of the accuracy vs. simplicity (interpretability) trade-off in the context in which it is applied.

This chapter reviews some of the most promising methods for supervised class prediction and highlights the potentialities and limitations of the single techniques, applied in different biological contexts.

In particular, two case studies are illustrated to elucidate the major differences between class prediction methods: the analysis of Khan *et al.* (2001) dataset about small round blue cell tumors (SRBCT) of childhood, and the analysis of expression data from the budding yeast *Saccharomyces cerevisiae* (Eisen *et al.*, 1998).

Nearest shrunken centroids (NSCs)

The method of nearest shrunken centroids (NSCs; Tibshirani *et al.*, 2002) identifies for each class the most characteristic subset of genes using a modified version of nearest-centroid method. The goal is to find the smallest set of genes that can accurately classify samples and could be the base for selecting and validate cancer markers. After selection of the minimal list of genes that are interesting for further investigations, the method searches for others genes that are highly correlated with the genes from the original list.

The algorithm starts by computing the centroid for each class, which is the average gene expression for each gene in each class.

The standard nearest-centroid method takes the gene expression profile of a new sample, and compares it to each of these class centroids. Then the class whose centroid is closest to the profile of the new sample is assigned to it. Nearest shrunken centroid classification differs from the standard nearest centroid method as it "shrinks" each of the class centroids toward the overall centroid for all classes by an amount called the "shrinkage threshold." This shrinkage consists of subtracting the threshold from the centroids, setting it equal to zero if it hits zero. For example, if the shrinkage threshold was 3.0, a centroid of 5.2 would be shrunk to 2.2, a centroid of –3.6 would be shrunk to –0.6, and a centroid of 2.1 would be set to zero. After applying the shrinkage, the new sample is assigned to the class of the closest centroid, as for the standard nearest centroid method. The use of a shrinkage threshold has the advantage of improving the accuracy of the classifier by removing noisy genes. For example, if after the shrinkage a gene has a centroid of zero for all classes, it is eliminated from the prediction.

The threshold is defined by minimizing the number of bad classified samples during the cross-validation. Basically the cross-validation consists of partitioning the training set into subsets and using just one of them for the training, while other subsets are used to validate the classifier.

The mathematical background of the NSC method is described in detail in the appendix section and in Tibshirani *et al.* (2002).

The software implementing NSC algorithm is available at http://www-stat.stanford.edu/~tibs/PAM/index.html. It uses the freely available R statistical package.

The NSC algorithm has been applied by Tibshirani *et al.* to the analysis of small round blue cell tumors (SRBCTs) of childhood (Khan *et al.*, 2001). The SRBCT dataset include neuroblastoma (NB), rhabdomyosarcoma (RMS), non-Hodgkin lymphoma (NHL) and the Ewing family of tumors. For the most appropriate form of treatment an accurate diagnosis is essential, but these cancers are difficult to distinguish by light microscopy. A gene expression based classification of SRBCTs could provide clinicians with an objective and highly accurate classification of tumors by simultaneous analysis of multiple markers. Originally, Khan *et al.* (2001) analyzed the datasets with a complex neural network approach and identified a subset of 96 genes needed for an accurate classification. Tibshirani *et al.*, applying the NSC algorithm to the same dataset, obtained better results—fewer genes that accurately predict classes. From this point of view NSC seems to outperform the neural networks presented by Khan *et al.* (2001). Fig. 6.1 shows the 10-fold cross-validation error as a function of the shrinkage threshold parameter Δ. We reran the analysis of Tibshirani *et al.* (2002) and we observed the minimal error rate for $\Delta = 4.41$. In this point only 44 genes are required to distinguish each class (with acceptable misclassification error level). If the number of genes is reduced to less than 44, the misclassification error increases. The list of the 44 selected genes is given in Fig. 6.2.

k-Nearest neighbor

The k-nearest neighbor (kNN) is based on the nearest-neighbor (NN) algorithm. Although the nearest-neighbor technique could be used without any *a priori* assignment (unsupervised), it has been applied successfully to classify biological objects based on their similarity to a known training set of both positive and negative cases.

In NN algorithm a gene expression profile is represented as an expression vector, so that each gene of the training set will be a vector in a multidimensional space. A new sample is then classified by computing the distance to each expression vector from the training set. The vector with the shortest distance is called the nearest neighbor, and its class determines the classification of the unknown sample.

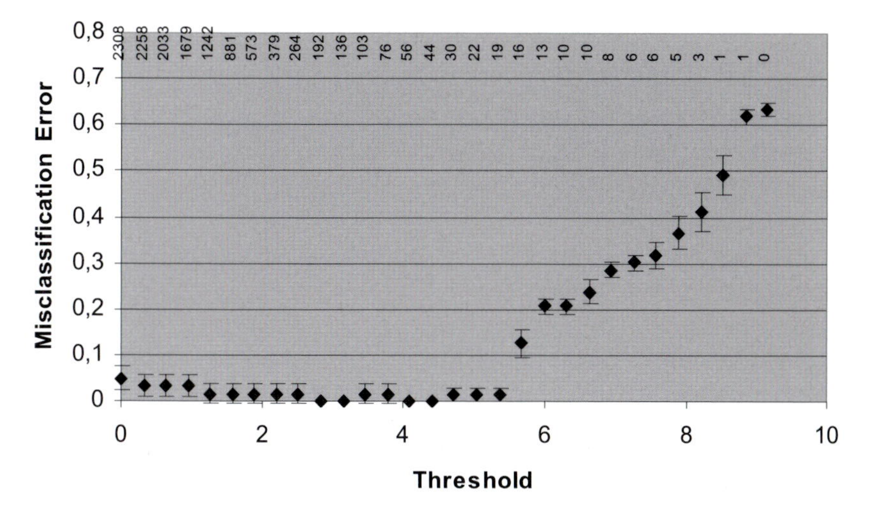

Figure 6.1 SBRCT classification: cross-validation error is shown as a function of the shrinkage threshold parameter Δ. The value $\Delta = 4.41$ yields a subset of 44 genes.

The kNN method further extends this approach by taking the *k* nearest points and assigning the class of the majority of them (Duda *et al.*, 2001). In practice it is a common rule to select an odd k to break ties. Increasing the *k* value will reduce the noisy within the training set, and as for the NSC method the optimal *k* is defined minimizing the cross-validation error rate. Further details of the algorithm are given in the appendix.

A disadvantage of this technique is the increasing computing power required as the size of the training set grows, since to classify an object the distance to each object in the training set needs to be computed. For large datasets this approach can be challenging. Several optimizations have been proposed, based on the computation of the pairwise distances using a restricted dataset.

This algorithm is implemented in GeneCluster 2.0 (Reich *et al.*, 2004), together with other unsupervised and supervised learning methods. In addition the software includes a permutation test to assess the statistical significance of the predicted classification. GeneCluster 2.0 is freely available at http://www.broad.mit.edu/cancer/software/genecluster2/gc2.html.

An analysis of leukemia subtypes described previously by Golub *et al.* provides a good example for a short description of the application. The distinction between acute lymphoblastic leukemia (ALL) and acute myeloid leukemia (AML) is crucial for effective treatment planning. Diagnosis is currently made by expert physicians on histopathological grounds. A classification based solely on gene expression profiles could help in defining a general strategy for discovering and predicting samples with accuracy. The set consists of 27 samples of ALL and 11 of AML. In their work, Golub *et al.* first built a predictor composed by a

770394	Fc fragment of IgG, receptor, transporter, alpha
207274	Human DNA for insulin-like growth factor II (IGF-2)
784224	fibroblast growth factor receptor 4
296448	insulin-like growth factor 2 (somatomedin A)
377461	caveolin 1, caveolae protein, 22kD
814260	follicular lymphoma variant translocation 1
241412	E74-like factor 1 (ets domain transcription factor)
244618	ESTs
183337	major histocompatibility complex, class II, DM alpha
1435862	antigen identified by mon. a/b 12E7, F21 and O13
812105	transmembrane protein
866702	protein tyrosine phosphatase, non-receptor type 13
295985	ESTs
796258	sarcoglycan, alpha
52076	olfactomedinrelated ER localized protein
325182	cadherin 2, N-cadherin (neuronal)
810057	cold shock domain protein A
80109	major histocompatibility complex, class II, DQ alpha 1
43733	glycogenin 2
357031	tumor necrosis factor, alpha-induced protein 6
298062	troponin T2, cardiac
767183	hematopoietic cell-specific Lyn substrate 1
1469292	pim-2 oncogene
461425	myosin MYL4
629896	microtubule-associated protein 1B
740604	interferon stimulated gene (20kD)
1409509	troponin T1, skeletal, slow
609663	protein kinase, cAMP-dependent, type II, beta
839552	nuclear receptor coactivator 1
236282	Wiskott-Aldrich syndrome
840942	major histocompatibility complex, class II, DP beta 1
898219	mesoderm specific transcript (mouse) homolog
1473131	transducin-like enhancer of split 2
42558	glycine amidinotransferase
47475	Homo sapiens inducible protein mRNA, complete cds
44563	Growth associated protein 43
841641	cyclin D1 (PRAD1: parathyroid adenomatosis 1)
769716	neurofibromin 2 (bilateral acoustic neuroma)
1471841	ATPase, Na+/K+ transporting, alpha 1 polypeptide
21652	Catenin (cadherin-associated protein), alpha 1 (102kD)
491565	Cbp/p300-interacting transactivator
814526	ESTs
143306	lymphocyte-specific protein 1
713922	glutathione S-transferase M1

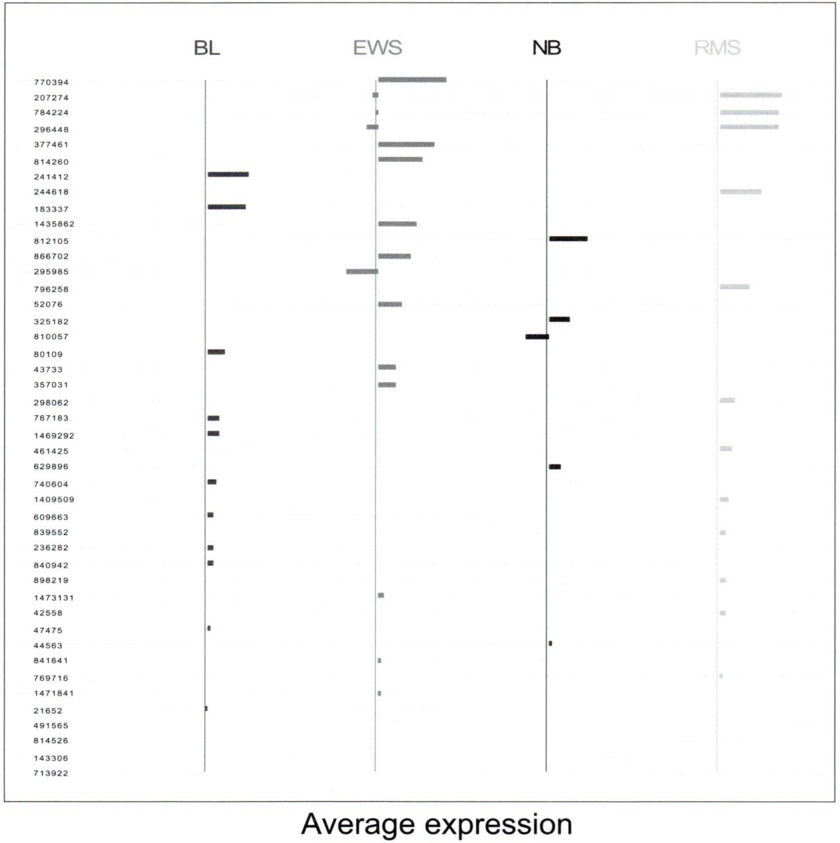

BL EWS NB RMS

Average expression

Figure 6.2 List and average expression value of selected 44 genes that are required to accurately distinguish SRBCT classes.

subset of the original 6,817 genes whose expression patterns strongly correlate with the two classes of leukemia. A gene with a strong discriminating power should have a high expression in one class and low in the other. They selected the best informative genes by cross-validation and used the first 50 of them to build a predictor. The 50-gene predictor assigned correctly 36 out of 38 samples using the kNN method, while the remaining two were defined as uncertain. One of the advantages of building a predictor is to reduce the noise associated to genes whose expression doesn't appear to be correlated with the groups, and drastically reduce the computational power by using a small subset of genes.

This strategy is implemented in GeneCluster, so we reran the analysis according to the Golub paper. The expression value of the best 50 predictive genes is shown in Fig. 6.3. Although genes appear globally correlated with a class, each single gene is not uniformly expressed across all the samples of a class. This emphasizes the interest in selecting a multigene predictor.

GeneCluster also offers the possibility of using a weighting scheme during the kNN assignment: each predictor gene votes for a class based on its expression level, and the weighting factor reflects how well the gene is correlated with the class distinction.

Support vector machine (SVM)

Support vector machines address the problem of discriminating groups of biological samples by means of combinations of a subset of genes (called features). In simple experiments it is easy to identify a small set of genes differentially expressed that correctly split the groups. Unfortunately, in complex experiments, where many more samples are profiled, is often impossible to identify a set of genes producing an accurate classification of the biological samples. SVMs expand the number of features by combining genes using mathematical functions.

SVM uses a training set to learn how to distinguish between members and non-members of a given class on the basis of expression data. In the training set biologically related genes are labeled positively, the rest is provided as negative examples. Each gene expression vector belonging to the training set can be regarded as a point in an n-dimensional space. As specified above, the optimal solution would be to define an hyperplane that successfully separates members of a class from non members. However, this plane could not exist for real-data. SVM circumvents this problem by increasing the number of dimensions using mathematical combinations of genes, named kernel functions: in this way even if two individual genes are not able to separate groups of samples, their combination could successfully separate them. In this way, creating a higher-dimensional space, any subset of samples can be separated. The risk of this approach is overfitting the training set, which could reduce the accuracy of prediction for unknown samples. SVM minimizes the overfitting by looking for a globally optimized solution, thus improving the predictive accuracy (Vapnik, 1998).

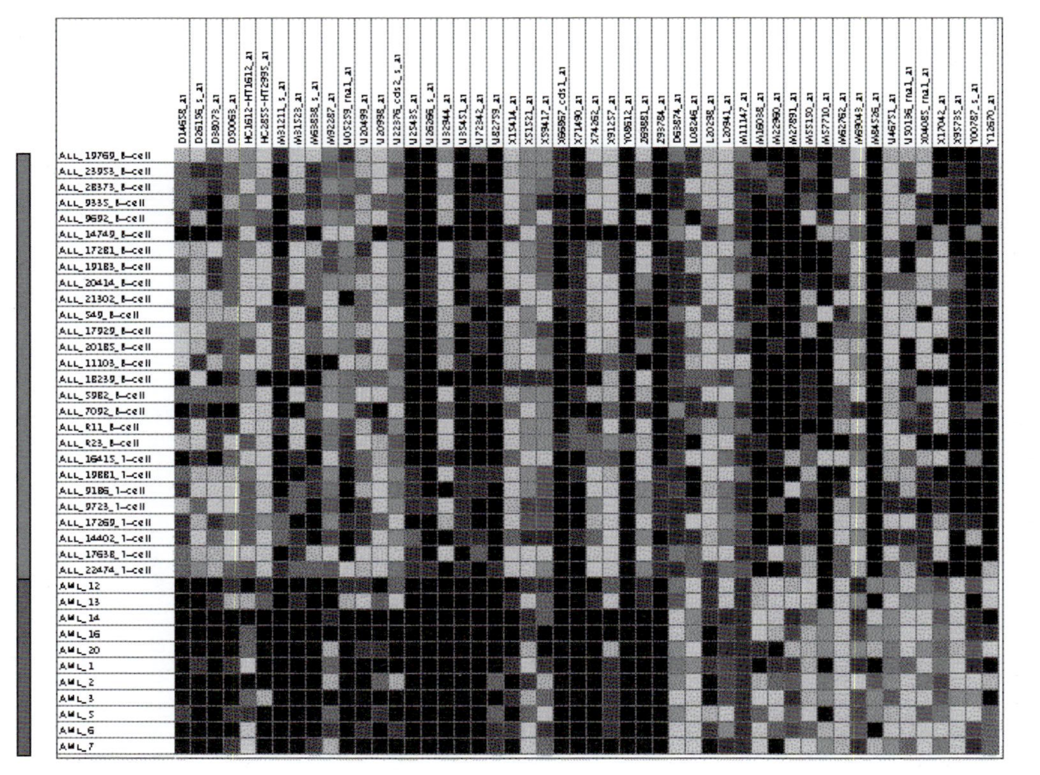

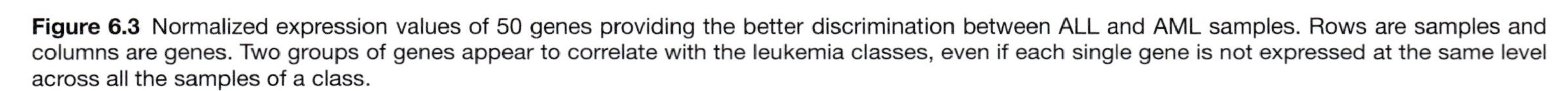

Figure 6.3 Normalized expression values of 50 genes providing the better discrimination between ALL and AML samples. Rows are samples and columns are genes. Two groups of genes appear to correlate with the leukemia classes, even if each single gene is not expressed at the same level across all the samples of a class.

Even if this strategy provides the most accurate solution from a mathematical point of view, the resulting plane is a mathematical combination of more genes, so its biological interpretation cannot be intuitive.

The SVM algorithm is defined so that it focuses on a small subset of the training set having the most important role in differentiating samples. This enables unknown samples to be efficiently classified, as the majority of the learning set can be discarded. Moreover, SVM is able to recognize outliers belonging to the training set or genes previously incorrectly assigned. More details about the algorithm are given in the appendix.

One of the softwares implementing SVM is *Genesis* (Sturn *et al.*, 2002), a Java suite for large-scale gene expression analysis. This program also includes several commonly used unsupervised and supervised clustering algorithms.

Five different supervised techniques, including SVM, were compared by Brown *et al.* (2000) using the classical Eisen dataset (Eisen *et al*, 1998) containing the expression of 2,467 genes across 79 time points. Six functional classes, defined according to the Munich Information Center for Protein Sequences (MIPS) catalogue (Mewes *et al.*, 2002), were used as training set. Five of them represent groups of genes that were expected to exhibit similar expression profiles, while the other one was included as control group. The results showed that SVM outperforms the other supervised methods for all classes and that annotations could be successfully inferred for several previously unannotated genes.

Supervised neural networks

Supervised neural networks (SNNs) are algorithms inspired by the structure and function of human neurons in brain.

Basically SNN implements a non-linear function of the form $Y(X, W)$, where Y is the output vector of the function given the input vector X and the network parameters W. In microarray classification a training set composed by n pairs $< X_i, Y_i >$, $i = 1, \ldots, n$, where X_i is a biological object and Y_i is its classification, is used to train the NN algorithm in order to determine the weight vector W that will best describe the training data.

During the training phase, in analogy to a biological learning, the vector W is determined by minimizing an error function that reflects the difference between our prediction and the known or expected prediction for the pair $< X_i, Y_i >$ from the training set. The obtained network is then validated with other available data and several performance measurements.

The major advantages of this approach are the non-linearity of the function Y(X, W) that allows better fit to the data, and its ability to deal with any type of data, thus expanding the knowledge available to train the algorithm. Given the ability of SNNs to solve problems in various domains, several efforts are currently dedicated to how to compute weights in order to optimize the fit.

One of the first neural networks was the perceptron (Rosenblatt, 1958), which established a mapping between input and output using a linear model having one single layer of weights. The perceptron initially seemed promising, but it was quickly proved that simple perceptrons could not be trained to accurately recognize many classes of patterns. Successively, more layers of trainable weights were added (multilayer perceptrons, MLPs), but these NNs still had a limited range of application due to the linearity of the model. The introduction of nonlinear capabilities further improved the algorithms, successfully predicting those cases where a linear discrimination failed. The learning procedure was also modified to enable a backward propagation (BP) of the error: in BP after a solution is produced, the corresponding error is computed (i.e., the difference between the expected classification vs. predicted) and backward propagated to modify the computed weights (Widrow and Lehr, 1990).

The presence of multiple layers, the nonlinear discrimination and the error propagation remarkably improved the noise tolerance and the robustness of neural networks thus improving the predictive accuracy.

Presently, different brands of NNs exist adapted to solve different situations. The Stuttgart Neural Network Simulator (SNNS) offers a flexible environment for research and applications on neural networks. It is freely available at the site: http://www-ra.informatik.uni-tuebingen.de/SNNS/.

Mateos *et al.* (2002) explored the ability of SNNs to learn gene expression patterns both in the binary and multiple class case using the Eisen dataset (Eisen *et al.*, 1998). Unlike Brown *et al.*, who analyzed the five classes expected to exhibit similar expression profiles, they investigated 96 classes of the Munich Information Center for Protein Sequences (MIPS) catalogue (Mewes *et al.*, 2002). What is interesting in the performed analysis is that 92% of functional classes gave > 60% false negatives. Moreover, some classes were learned with a low rate of false positives, others hardly learn at all. Three reasons for poor learning performance were identified: class size, heterogeneity of the classes and high degree of intersection among functional classes. Classes of larger sizes learned more efficiently than the small size classes. A second factor—heterogeneity—also made learning difficult, if the class members are very different from each other. The reason for the high interconnectivity of MIPS catalogue is that cellular processes do not represent isolated parts of live machinery. This means that it is not necessarily a failure of the learning scheme if the neural networks assign a gene to different classes. Mateos *et al.* also introduced two numerical indices that gave a simple measure of the overlap between classes: the link number per gene and the relative sharing number.

Simple comparison of supervised neural networks (SNN, applied as multilayer perceptron—MLP) and support vector machine (SVM) done by Mateos *et al.* shows that MLP and SVM perform similarly in terms of false-negatives (FNs), although the MLP seems to do systematically worse compared with SVM in terms of the number of false-positives (FPs). The inferiority of MLP to SVM is observed despite the MLP ability to perform multiclass classification, but as discussed in the above section, FPs and FNs corresponding to misclassified samples might not reflect true biological FPs and FNs.

Random forests (RFs)

Random forests (Breiman, 2001) are an extension of the more general method named "classification trees" or "decision trees." In general, the objective of a decision tree is to build a predictor that, as for the other supervised learning approaches, could predict a response based on an input variable. The main difference from other predictive classification methods is the hierarchical nature of the results and the flexibility of the method.

In RF the strategy used to examine the effects of the predictor variables on the class assignment is hierarchical: the variables are recursively analyzed on one by one basis, and a binary split is defined for each variable.

As an example, we can suppose setting up a classification tree to discriminate between coins based on their diameters. If the diameter is more than say 10, the coin is assigned to the class *a*, otherwise the algorithm will proceed to the next step. If the coin has a diameter > 7, it is assigned to the class *b*, otherwise the algorithm will continue to the next step, and so on. This recursive approach results in a classification tree.

As variable are processed independently, different types of predictors can be analyzed (i.e., categorical, such as type of disease, and continuous, such as blood pressure). Finally, unnecessary leaves and branches are removed to improve the performance of the algorithm.

The possibility of adding knowledge deriving from different sources clearly increases the flexibility of this approach that can be applied to a wide variety of classification problems. The drawback is that more traditional methods based on certain distributional assumptions may perform better in certain specific cases. However, classification trees represent an interesting tool for exploratory data analysis. In addition, given the hierarchical structure of the algorithm, a classification tree can be translated into a series of rules easily understandable (i.e., if the coin has a diameter < 10 and > 7, it is assigned to the class *b*).

RF expands the concept behind the classification tree by assigning new objects on the basis of different classification trees. The class prediction for the object is based on the agreement between all the trees in the forest. A random forest can be generated in two different ways: (i) by extracting several bootstrap samples from the original dataset and building a tree for each bootstrap replicate; (ii) by random splitting nodes in a tree. In both cases a forest will be generated. Although these approaches will improve the accuracy of the classification, the random element introduced to generate the forest implies that results lack complete reproducibility. To circumvent this problem an alternative methodology has been proposed by Zhang *et al.* (2003) using a different resampling procedure to generate the forest.

An interesting example showing the application of RF to microarray data analysis is represented by the work of Gunther *et al.* (2003). They have used RF to successfully predict drug efficacy, on the basis of gene expression profiles induced by the drugs (Gunther *et al.*, 2003).

Conclusions

We have shown how methods of artificial intelligence can be successfully used to classify biological samples and genes on the basis of transcriptional data.

The tools and techniques described here are not meant to be comprehensive and there are new tools and algorithms under development. We have illustrated the methods that have been so far successfully used; although some algorithm can perform better in specific situations, it must be emphasized that the evaluation of the "right" algorithm is driven by our understanding of the biological problem under analysis. A classification algorithm is considered valid if, besides producing novel perspectives for data interpretation, the results are compatible with our previous knowledge and experimental experience in that research area.

The supervised classification techniques we illustrated, as well as the new ones that will be developed during the coming years, promise to be crucial for the elucidation of mechanisms underlying many biological processes, most importantly those leading to a diseased phenotype. As recent studies demonstrate, we are not far from being able to use computer based classification approaches as customary tools for diagnosis and prognosis assessment in important clinical area such as cancer. As these techniques are based on exhaustive molecular data, potentially representing the entire "transcriptome," the level of resolution that can be reached is intrinsically much higher than traditional classification and prognostic methods, based on morphological observations and/or low throughput biochemical assays.

Supervised classification methods also offer the exciting opportunity to explore hidden relationships between genes, often not apparent in unsupervised, exploratory analyses. The application of these approaches to microarray cancer studies has shown to be effective in the identification of genes whose altered expression can be used as biomarkers.

Acknowledgements

This work was supported by EC BioSapiens (LHSG-CT-2003-503265) 6FP project as well as the Polish Ministry of Education and Science (PBZ-MNil-2/1/2005 and 2P05A00130). The authors would like to acknowledge Janet Clench for the English revision of the text.

Appendix A: Classification algorithms

Support vector machine (SVM)

SVM is an effective statistical learning method (Vapnik, 1995, 1998; Cristianini and Shawe-Taylor, 2000) with good performance yet easier to implement than neural networks. It was successfully applied to various problems including text classification (Joachims, 2002), image recognition tasks (Vojtech, 2003), bioinformatics (Kim and Park, 2003; Minakuchi *et al.*, 2003) and medical applications

(Guyon *et al.*, 2002; Valentini, 2002). The SVM approach has been used also in analysis of gene expression data (Brown *et al.*, 2000), classification of microarrays data (Furey *et al.*, 2000), to infer gene functional classification (Krishnan and Westhead, 2003; Pavlidis, 2001; Zien *et al.*, 2000) and for protein analysis (Ding and Dubchak, 2001; Hua and Sun, 2001; Jaakkola *et al.*, 2000).

The output of the training phase is a classification function, i.e., a model. It consists of a set of D support vectors T_j and α_i, which are nonzero, positive real numbers. Those constants are obtained from optimization procedure (quadratic programming QP problem) used to find the maximal margin hyperplane. The number of free parameters of the QP problem is equal to the number of all instances in the training dataset. The non-zero parameters α_i describe the strength of this particular *i*th support vector in the decision function. SVM chooses as support vectors those points that lie closest to the separating hyperplane. The kernel function is used to define the feature space after nonlinear mapping function from the embedding space. The mapping function Ω need not be explicitly defined because only the inner product in the kernel function is used. The kernel function is a positive define function reflecting the similarity between an input sample and the set of support vectors T_i. In most cases three types of kernels are used: the linear, polynomial or radial basis.

The reliability of a classification of a single segment (Schölkopf *et al.*, 1999) as a functional site is given by the cost function:

$$f(T[x]) = \sum_{i=1}^{i=D} l_i \alpha_i K(\Omega\{T[x]\}, \Omega\{T_i\})$$

where $K(T, T_i)$ is the proper kernel function that defines the feature space, Ω is a nonlinear mapping function from embedding space T into the feature space, and l_i are known *a priori* class labels for support vectors. We use $l_i = +1$ for positive cases and $l_i = -1$ for negative ones. The kernel function is a positive define function reflecting the similarity between an input sample and the set of support vectors T_i. The non-zero parameters α_i describe the strength of this particular *i*th support vector in the decision function. SVM chooses as support vectors those points that lie closest to the separating hyperplane.

Nearest shrunken centroids (NSCs)

The method was previously described by Tibshirani *et al.* (2002). Let x_{ij} the expression for genes $i = 1, 2, 3 \ldots p$ and samples $j = 1, 2, 3 \ldots n$. We have classes $1, 2, \ldots K$, and let C_k be indices of the n_k samples in class k. The *i*th component of the centroid for class k is:

$$\bar{x}_{ik} = \sum_{j \in C_k} \frac{x_{ij}}{n_k} \tag{6.1}$$

the mean expression value in class k for gene i; the *i*th component of the overall centroid is:

$$\bar{x}_i = \sum_{j=1}^{n} \frac{x_{ij}}{n} \tag{6.2}$$

Let

$$d_{ik} = \frac{\bar{x}_{ik} - \bar{x}_i}{m_k \cdot (s_i + s_0)}, \tag{6.3}$$

where s_i is the pooled within-class standard deviation for gene i:

$$s_i^2 = \frac{1}{n-K} \sum_k \sum_{j \in C_k} (x_{ij} - \bar{x}_{ik})^2 \tag{6.4}$$

and

$$m_k = \sqrt{\frac{1}{n_k} + \frac{1}{n}} \tag{6.5}$$

makes the $m_k \cdot s_i$ equal to the estimated standard error of the numerator in d_{ik}. In the denominator, the value s_0 is a positive constant (with the same value for all genes), included to guard against the possibility of large d_{ik} values arising by chance from genes with low expression levels. We set s_0 equal to the median value of the s_i over the set of genes.

Thus, d_{ik} is a t statistic for gene i, comparing class k to the overall centroid. We rewrite Eqn (6.3) as:

$$\bar{x}_{ik} = \bar{x}_i + m_k (s_i + s_0) d_{ik} \tag{6.6}$$

The NSC method shrinks each d_{ik} toward zero, giving d_{ik} and yielding shrunken centroids or prototypes:

$$\bar{x}'_{ik} = \bar{x}_i + m_k (s_i + s_0) d'_{ik} \tag{6.7}$$

This shrinkage is called soft thresholding: each d_{ik} is reduced by an amount Δ in absolute value and is set to zero if its absolute value is less than zero. Algebraically, soft thresholding is defined by:

$$d'_{ik} = sign(d_{ik})(|d_{ik}| - \Delta)_+ \tag{6.8}$$

where + means positive part ($t_+ = t$ if $t > 0$ and zero otherwise).

k-Nearest neighbor (kNN)

The kNN-based method selects genes with expression profiles similar to the gene of interest to impute missing values. If we consider gene A that has one missing value in the first experiment, this method finds k other genes, which have a value present in the first experiment, with expression most similar to A in experiments 2 *M* (where *M* is the total number of experiments). A weighted average of values in experiment *1* from the *k* closest genes is then used as an estimate for the missing value in gene *A*. In the weighted average, the contribution of each gene is weighted by similarity of its expression to that of gene *A*. In most cases the similarity is computed using the Euclidean distance. The Euclidean distance measure is often sensitive to outliers, which could be present in microarray data. However the log-transforming the data reduces the effect of outliers on gene similarity determination (Troyanskaya *et al.*, 2001).

To be more precise in the KNN method, one computes the distance *D* between a sample, represented by its pattern vector V_m, and each of the pattern vectors of the training set:

$$V_m = (g_{1m} \cdots g_{im} \cdots g_{nm}) \tag{6.9}$$

where *n* is the number of genes in the vector; g_{im} is the expression level (log10 transformed) of *i*th gene in the *m*th sample; $m = 1, \ldots, M$. Each sample is classified according to the class membership of its *k* nearest neighbors (if they agree), as determined by the Euclidean distance in *n*-dimensional space. For most cases small values of *k* have been proven to provide good classification results, one should select *k* value large enough to form tight clusters even if there are subtypes and the sample size is limited (Li *et al.*, 2001). In some versions of the kNN method the class membership is determined by majority vote of the *k*-nearest neighbors. If majority of the *k*-nearest neighbors of a sample are tumors, the sample is classified as tumor. A sample remains unclassified if the number of nearest neighbors belonging to the same class doesn't exceed a fixed threshold.

In order to estimate the missing value g_{im} of gene *i* in sample *m*, one should select *k* genes that have expression vectors most similar to genetic expression of gene *i* in samples other than *m*. The similarity measure between two expression vectors V_n and V_m is determined by the simple Euclidian distance D_{nm} over the observed components in sample *m*:

$$D_{nm} = |V_n - V_m| \tag{6.10}$$

The missing value is then estimated as the weighted average of the corresponding entries in the selected *k* expression vectors:

$$\hat{g}_{im} = \sum_{i=1}^{k} W_i \cdot X_i \tag{6.11}$$

with

$$W_i = \frac{1}{D_i \cdot \Delta} \tag{6.12}$$

where

$$\Delta = \sum_{i=1}^{k} D_i \quad (6.13)$$

D_i and X is the input matrix containing gene expressions. Those equations show that each gene contribution is weighted by the similarity of its expression to gene i (Sehgal *et al.*, 2005).

The kNN method is robust enough to increase the fraction of data missing and shows less deterioration in performance with increasing percent of missing entries (Troyanskaya *et al.*, 2001). In addition, the kNN method is very robust to the type of data for which estimation is performed, performing better on non-time series or noisy data and also less sensitive to the exact parameters used (for example the number of nearest neighbors). kNN has the advantage of providing accurate estimation for missing values in genes that belong to small tight expression clusters. Missing points for such genes could be estimated poorly by other methods (for example SVD-based estimation) if their expression pattern is not similar to any of the eigengenes used for regression (Troyanskaya *et al.*, 2001).

The computational complexity of kNN is $O(m^2n)$, where m and n are the number of genes and samples. A vital feature of KNN is that it does not consider negative correlations between data, which can lead to estimation errors. kNN-based imputation provides a robust and sensitive approach to estimating missing data for microarrays. The goal of this algorithm is to provide an accurate way of estimating missing values in order to minimally bias the performance of microarray analysis methods. However, its significance to biological discovery should be always assessed in order to avoid false conclusions.

Appendix B: Analysis software systems

Gene Cluster 2.0

GeneCluster 2.0 implements supervised classification, gene selection and permutation test methods. It also includes algorithms for building and testing supervised models using weighted voting and k-nearest neighbors (kNN) algorithms, module for batch SOM clustering, a marker gene finder based on a kNN analysis and a visualization module.

Web address: http://www.broad.mit.edu/cancer/software/genecluster2/gc2.html
Interface/operating system: Windows, Mac, Unix

PAM

PAM (Prediction Analysis of Microarrays) is a statistical technique for class prediction from gene expression data using nearest shrunken centroids. It is described in Tibshirani *et al.* (2002). The technique is general and can be used in many other classification problems. It can also be applied to survival analysis problems. PAM is available as a library for the R package.

Web address: http://www-stat.stanford.edu/~tibs/PAM/index.htm
Interface/operating system: Windows, Linux

Genesis

Genesis is a Java suite for large-scale gene expression analysis. The main features are data filtering, data normalization, clustering (including hierarchical clustering, k-means, SOM), principal component analysis, correspondence analysis, one-way ANOVA for detection of differentially expressed genes, support-vector machine, integration of the Gene Ontology.

Web address: http://genome.tugraz.at/Software/GenesisCenter.htm
Interface/operating system: Windows, LINUX, UNIX, Solaris and Irix

Stuttgart Neural Network Simulator (SNNS)

SNNS is a software simulator for neural networks developed at the Institute for Parallel and Distributed High Performance Systems (IPVR) at the University of Stuttgart. The goal of the SNNS project is to create an efficient and flexible simulation environment for research on and application of neural nets.

Web address: http://www-ra.informatik.uni-tuebingen.de/SNNS/
Interface/operating system: LINUX, SUN, Windows

Random Forest

Random Forest is an R package implementing the Breiman and Cutler's random forests for classification and regression.

Web address: http://cran.r-project.org/src/contrib/Descriptions/randomForest.html
Interface/operating system: LINUX, Windows

References

Alizadeh, A.A., Eisen, M.B., Davis, R.E., Ma, C., Lossos, I.S., Rosenwald, A., Boldrick, J.C., Sabet, H., Tran, T., Yu, X., *et al.* (2000). Distinct types of diffuse large B-cell lymphoma identified by gene expression profiling. Nature 403, 503–511.

Bittner, M., Meltzer, P., Chen, Y., Jiang, Y., Seftor, E., Hendrix, M., Radmacher, M., Simon, R., Yakhini, Z., Ben-Dor, A., *et al.* (2000). Molecular classification of cutaneous malignant melanoma by gene expression profiling. Nature 406, 536–540.

Breiman. L. (2001). Mach. Learn. 45, 5–32

Brown, M.P., Grundy, W.N., Lin, D., Cristianini, N., Sugnet, C.W., Furey, T.S., Ares, M., Jr., and Haussler, D. (2000). Knowledge-based analysis of microarray gene expression data by using support vector machines. Proc. Natl. Acad. Sci. USA 97, 262–267.

Cristianini, N., and Shawe-Taylor, J. (2000). An introduction to support vector machines: and other kernel-based learning methods (Cambridge, U.K.; New York: Cambridge University Press).

Ding, C.H., and Dubchak, I. (2001). Multi-class protein fold recognition using support vector machines and neural networks. Bioinformatics *17*, 349–358.

Duda, R.O., Hart, P.E., and Stork, D.G. (2001). Pattern classification, 2nd edn (New York: Wiley).

Eisen, M.B., Spellman, P.T., Brown, P.O., and Botstein, D. (1998). Cluster analysis and display of genome-wide expression patterns. Proc. Natl. Acad. Sci. USA *95*, 14863–14868.

Furey, T.S., Cristianini, N., Duffy, N., Bednarski, D. W., Schummer, M., and Haussler, D. (2000). Support vector machine classification and validation of cancer tissue samples using microarray expression data. Bioinformatics *16*, 906–914.

Golub, T.R., Slonim, D.K., Tamayo, P., Huard, C., Gaasenbeek, M., Mesirov, J. P., Coller, H., Loh, M.L., Downing, J. R., Caligiuri, M. A., *et al.* (1999). Molecular classification of cancer: class discovery and class prediction by gene expression monitoring. Science *286*, 531–537.

Gunther, E.C., Stone, D.J., Gerwien, R.W., Bento, P. and Heyes, M.P. (2003). Prediction of clinical drug efficacy by classification of drug-induced genomic expression profiles in vitro. Proc. Natl. Acad. Sci. USA *100*, 9608–9613

Guyon, I., Weston, J., Barnhill, S., and Vapnik, V. (2002). Gene selection for cancer classification using support vector machines. Mach. Learn. *46*, 389—422.

Hedenfalk, I., Duggan, D., Chen, Y., Radmacher, M., Bittner, M., Simon, R., Meltzer, P., Gusterson, B., Esteller, M., Kallioniemi, O.P., *et al.* (2001). Gene-expression profiles in hereditary breast cancer. N. Engl. J. Med. *344*, 539–548.

Hua, S., and Sun, Z. (2001). A novel method of protein secondary structure prediction with high segment overlap measure: support vector machine approach. J. Mol. Biol. *308*, 397–407.

Jaakkola, T., Diekhans, M., and Haussler, D. (2000). A discriminative framework for detecting remote protein homologies. J. Comput. Biol. *7*, 95–114.

Joachims, T. (2002). Learning to classify text using support vector machines (Boston: Kluwer Academic Publishers).

Khan, J., Wei, J.S., Ringner, M., Saal, L.H., Ladanyi, M., Westermann, F., Berthold, F., Schwab, M., Antonescu, C. R., Peterson, C., and Meltzer, P.S. (2001). Classification and diagnostic prediction of cancers using gene expression profiling and artificial neural networks. Nat. Med. *7*, 673–679.

Kim, H., and Park, H. (2003). Protein secondary structure prediction based on an improved support vector machines approach. Protein Eng. *16*, 553–560.

Krishnan, V.G., and Westhead, D.R. (2003). A comparative study of machine-learning methods to predict the effects of single nucleotide polymorphisms on protein function. Bioinformatics *19*, 2199–2209.

Li, L., Weinberg, C.R., Darden, T.A., and Pedersen, L.G. (2001). Gene selection for sample classification based on gene expression data: study of sensitivity to choice of parameters of the GA/KNN method. Bioinformatics *17*, 1131–1142.

Mateos, A., Dopazo, J., Jansen, R., Tu, Y., Gerstein, M., and Stolovitzky, G. (2002). Systematic learning of gene functional classes from DNA array expression data by using multilayer perceptrons. Genome Res. *12*, 1703–1715.

Mewes, H. W., Frishman, D., Guldener, U., Mannhaupt, G., Mayer, K., Mokrejs, M., Morgenstern, B., Munsterkotter, M., Rudd, S., and Weil, B. (2002). MIPS: a database for genomes and protein sequences. Nucleic Acids Res. *30*, 31–34.

Minakuchi, Y., Satou, K., Konagaya, A. (2003). Prediction of protein-protein interaction sites using supprot vector machnes, Paper presented at: International conference on mathematics and engineering techniques in medicine and biological sciences.

Pavlidis, P., Weston, J., Cai, J., Grundy, W.N. (2001). Gene functional classification from heterogeneous data, Paper presented at: 5th International Conference on Computational Molecular Biology (NY, Montreal, Canada: ACM Press).

Reich, M., Ohm, K., Angelo, M., Tamayo, P., and Mesirov, J P. (2004). GeneCluster 2.0: an advanced toolset for bioarray analysis. Bioinformatics *20*, 1797–1798.

Rosenblatt, F. (1958). The perceptron: a probabilistic model for information storage and organization in the brain. Psychol. Rev. *65*, 386–408.

Schölkopf, B., Burges, C.J.C., and Smola, A.. (1999). Advances in kernel methods: support vector learning (Cambridge, MA: MIT Press).

Sehgal, M. S., Gondal, I., and Dooley, L S. (2005). Collateral missing value imputation: a new robust missing value estimation algorithm for microarray data. Bioinformatics *21*, 2417–2423.

Sturn, A., Quackenbush, J., and Trajanoski, Z. (2002). Genesis: cluster analysis of microarray data. Bioinformatics *18*, 207–208.

Tibshirani, R., Hastie, T., Narasimhan, B., and Chu, G. (2002). Diagnosis of multiple cancer types by shrunken centroids of gene expression. Proc. Natl. Acad. Sci. USA *99*, 6567–6572.

Troyanskaya, O., Cantor, M., Sherlock, G., Brown, P., Hastie, T., Tibshirani, R., Botstein, D., and Altman, R. B. (2001). Missing value estimation methods for DNA microarrays. Bioinformatics *17*, 520–525.

Valentini, G. (2002). Gene expression data analysis of human lymphoma using support vector machines and output coding ensembles. Artificial Intelligence in Medicine *26*, 281–304.

Vapnik, V.N. (1995). The nature of statistical learning theory (New York: Springer).

Vapnik, V.N. (1998). Statistical learning theory (New York: Wiley).

Vojtech, F., Vaclay, H. (2003). An iterative algorithm learning the maximal margin classifier. Pattern Recog. *36*, 1985–1996.

Widrow. B. and Lehr, M. A. (1990). 30 Years of Adaptative Neural Networks: Perceptron, Madaline and Backpropagation, Paper presented at: IEEE (78).

Yeang, C.H., Ramaswamy, S., Tamayo, P., Mukherjee, S., Rifkin, R. M., Angelo, M., Reich, M., Lander, E., Mesirov, J., and Golub, T. (2001). Molecular classification of multiple tumor types. Bioinformatics *17 Suppl 1*, S316–322.

Zavaljevski, N., Stevens, F. J., and Reifman, J. (2002). Support vector machines with selective kernel scaling for protein classification and identification of key amino acid positions. Bioinformatics *18*, 689–696.

Zhang, H., Yu, C.Y., and Singer, B. (2003). Cell and tumor classification using gene expression data: Construction of forests. Proc. Natl. Acad. Sci. USA *100*, 4168–4172.

Zien, A., Ratsch, G., Mika, S., Scholkopf, B., Lengauer, T., and Muller, K.R. (2000). Engineering support vector machine kernels that recognize translation initiation sites. Bioinformatics *16*, 799–807.

A Case Study: The Mammary Carcinogenesis in HER2 Transgenic Mice

7

Federica Cavallo, Guido Forni, Anna Grassi, PierLuigi Lollini, and Raffaele Calogero

Abstract

Microarray transcription profiling was applied on BALB-neuT breast cancer model to understand the molecular mechanisms associated to the halting of tumor growth via HER2 vaccination. The demonstration that vaccines can cure HER2 transplantable tumors was achieved both via vaccination with allogeneic mammary carcinoma cells expressing high levels of both r-HER2 and H-2^q class I molecules followed by administration of IL-12 or via conventional intramuscular DNA vaccination followed by boost with H-2^q r-HER2 positive tumor cells gene-engineered to release interferon-γ (IFN-γ). The combination of transcription profiles with protection results indicated that the main effect was a strong polyclonal antibody response, and chronic vaccination is needed to maintain an active IFN-γ-mediated response in the mammary gland. Furthermore, cross-study comparison of BALB-neuT gene expression array data opens the way to the identification of new tumor-associated antigens (TAAs) to be used in conjunction with HER2 to allow a broader coverage of vaccination.

Introduction

Microarray data analysis is a technology that might give rise to both strong satisfaction and strong frustration to the users. Microarray analysis can be used as an exploratory or a confirmatory tool. In the first case transcription profiling is used to generate set of data that, after validation with alternative methods, will be used to generate new hypothesis to be investigated. This approach could be very useful to define new interesting research path, however those results will need long bench work before reaching publishable conclusions. The use of transcription profiles as a confirmatory tool is instead a valuable tool to give more strength to already available biological evidences.

The test case we are presenting in this chapter represents an example of the usage of the microarray analysis as confirmatory and exploratory tool. The present paper will focus on the BALB-neuT breast cancer model. We will present the results obtained by independent groups (Quaglino *et al.*, 2004; Astolfi *et al.*,

2005a) on the usage of microarray as confirmatory approach of immunological/histological evidences related to the mechanism of action of HER2-driven vaccination in halting tumor growth. On the other side, part of the microarray data generated in these two studies were also used to investigate the possibility to identify, by transcriptional analysis, new TAAs to be used in conjunction with HER2 to improve the vaccination coverage.

The HER2 oncogene

Tyrosine kinase receptors of the HER (human EGF receptor) family form a signaling network that fulfils fundamental functions during development and post-natal life (Yarden and Sliwkowski, 2001). The HER2 receptor is central to the network's function in mammalian cells. HER2 does not bind ligands directly, but acts as a co-receptor for several ligands in the context of heterodimers with other HER family members (Klapper *et al.*, 2000). Hence, assembly of HER2 in receptor complexes endows them with a remarkable signaling superiority in the form of a more intense and more prolonged signal output due to the ability of HER2 to impose slower ligand off-rates and slower rates of receptor internalization/degradation on receptor dimers. Receptor signals are thus propagated at a higher intensity and for a longer time, an effect which is even more pronounced upon HER2 over-expression and also leads to ligand-independent constitutive activation of HER2 homodimers (Mosesson and Yarden, 2004).

Amplification and consequent over-expression of the HER2 gene occur in 20–30% of breast cancers (Mosesson and Yarden, 2004) and are accompanied by a more aggressive disease course and a poorer prognosis (Pupa *et al.*, 2005). In addition to this epidemiological correlation, the picture of HER2-driven breast cancer in transgenic mice (Lollini and Forni, 2003) provides cogent evidence in favor of the view that HER2 over-expression may be a critical element in the pathogenesis of the human form.

The BALB-neuT breast carcinoma murine model

We have generated a syngeneic (H-2^d) strain of mice transgenic for the transforming rat (r-) *HER2* oncogene (BALB-neuT mice) (http://cancermodels.nci.nih.gov/mmhcc/index.jsp), starting from a non-inbred male transgenic for this oncogene (Lucchini *et al.*, 1992) under the transcriptional control of the mammary tumor virus promoter (Muller *et al.*, 1988). BALB-neuT virgin females inexorably display a palpable invasive carcinoma in all their ten mammary glands around the 33rd week of age. The progression of these lesions is consistent. Around the nipple, epithelial nodular neoformations (side buds) stemming from the main and secondary mammary ducts are already evident in 4-week-old mice (Di Carlo *et al.*, 1999). Histologically, these buds are the foci of atypical hyperplasia in carcinomatous progression. At 8 weeks, most side buds have progressed to *in situ* carcinomas, while an increasing number of hyperplastic foci are evident all over the gland. Between the 10th and the 20th week, the *in situ* carcinomas metastasize to the bone marrow and lungs. This progression is similar in all ten glands. The cells of the hyperplastic, neoplastic and metastatic lesions constantly highly overexpress the r-HER2 receptor in both their cytoplasm and their membrane (Boggio *et al.*, 1998; Di Carlo *et al.*, 1999).

Halting BALB-neuT carcinoma by HER2 vaccination

The association of the overexpression of HER2 with poor prognosis has led to recognition of the therapeutic potential of its targeting by means of drugs and in immunopreventive strategies. The progression of mammary lesions in BALB-neuT mice provides an unique experimental setting in which to assess the ability of drugs and immunostimulating maneuvers to inhibit their HER2-driven carcinogenesis and determine what is needed to maintain protection in the hope that the results can be employed for the successful long-term management of human HER2-driven tumors. Mammary tumor progression is altered when a specific immunity to HER2 is elicited in BALB-neuT mice (Spadaro *et al.*, 2004). Vaccination with allogeneic mammary carcinoma cells expressing high levels of both r-HER2 and H-2^q class I molecules followed by administration of IL-12 effectively and persistently halts carcinogenesis (Nanni *et al.*, 2001).

Beginning at 6 weeks of age, BALB-neuT mice received IL-12 twice a week in the first and second week, followed by five daily administrations in the third. After one week of rest, this course was repeated until mice were sacrificed or reached 52 weeks. This chronic combined treatment, called "Triplex" vaccination, rendered all mice tumor-free at week 33. At 52 weeks, 88% were still completely tumor-free. Almost the same protection was provided by a single-cell vaccine composed of H-2^q r-HER2 positive tumor cells gene-engineered to release IL-12 (De Giovanni *et al.*, 2004). These findings show that initial preneoplastic lesions are an appropriate and rational target for a specific immunological attack.

To determine whether immunity can also hamper the progression of more advanced lesions, BALB-neuT mice bearing multiple *in situ* carcinomas were vaccinated (primed) at week 10 and 12 with DNA plasmids coding for the extracellular and transmembrane domain of r-HER2, but lacking the intracellular kinase domain in order to endow the recipient cells with membrane expression without any risk of neoplastic transformation. We used a conventional intramuscular DNA vaccination. Primed mice were boosted at week 13 with H-2^q r-HER2 positive tumor cells gene-engineered to release interferon-γ (Quaglino *et al.*, 2004). This treatment halts carcinogenesis, though inhibited neoplastic lesions still persist and eventually give rise to lethal carcinomas (Lo Iacono *et al.*, 2005).

Despite the differences in the treatment schedule and vaccine composition of these studies, the results are remarkably convergent. Sequential histology and whole-mount show that, as long as the protection lasts, the mammary lesions present at the time of vaccination remain unchanged. Moreover, both studies suggest that this efficient inhibition rests on the concurrence of anti-HER2 antibodies and T-cell-mediated delayed hypersensitivity (Quaglino *et al.*, 2004). These data are in contrast with the overwhelming evidence that eradication of fast-growing transplantable tumors rests on swift high-affinity T-cell reactivity, whereas B-cells may interfere with the efficiency of the reaction. To further explore this issue, the gene expression patterns of the mammary glands of vaccinated BALB-neuT mice have been investigated by DNA microarray analysis (Quaglino *et al.*, 2004; Astolfi *et al.*, 2005a) and the results have been compared (Astolfi *et al.*, 2005b).

Vaccination effects and transcriptional profiling

A major problem in tumor immunology is evaluation of the immunological response during immunotherapy. There are thousands of tests potentially relevant to tumor immunity, ranging from antigen expression to cytokines and chemokines, and cellular and humoral activities, but their application in every case is obviously impossible. One solution is offered by DNA microarrays, which simultaneously measure the transcription level of thousands of genes up to the entire transcriptome of any given cell or tissue. Investigators are not confined to a small set of predefined genes and examination of the transcriptome can reveal hidden pathways activated or inhibited by immune responses and by immunotherapy.

The feasibility of this approach has been assessed in a complete microarray analysis of the genes activated in the mammary glands of BALB-neuT mice during progression from hyperplasia to carcinoma, and in mice receiving Triplex (Astolfi *et al.*, 2005a) or prime-and-boost (Quaglino *et al.* 2004) vaccination.

Triplex vaccination

DNA microarrays (27 MG-U74Av2 Affymetrix GeneChips) were used to analyze the arrest of tumorigenesis induced in BALB-neuT mice by Triplex vaccination (Astolfi *et al.*, 2005a). Vaccination commenced at the 6th week of age and was repeated lifelong. The gene expression profile of the mammary tissue of not-treated BALB-neuT mice was analyzed at 6, 15, 19, and 26 weeks of age (wk6nt, wk15nt, wk19nt, and wk26nt respectively), and compared with that of mammary tissue of 15- and 26-week-old vaccinated mice (wk15vax, and wk26vax respectively). Probe set intensities were background-corrected, normalized, and summarized by the Robust Multi-Array analysis (RMA) method (Irizarry *et al.*, 2003a,b) implemented in the Affy package of Bioconductor (Gentleman *et al.*, 2004).

PCA (Raychaudhuri *et al.*, 2000) was used to investigate the overall performance of the vaccination. This technique represents the global gene expression changes between the samples as coordinates in a multidimensional space in which the axes are called the "principal components." Since in this experiment, the first two components account for the greatest part of the variability within the dataset (70.2%), their graphical representation (Fig. 7.1) will catch the differences

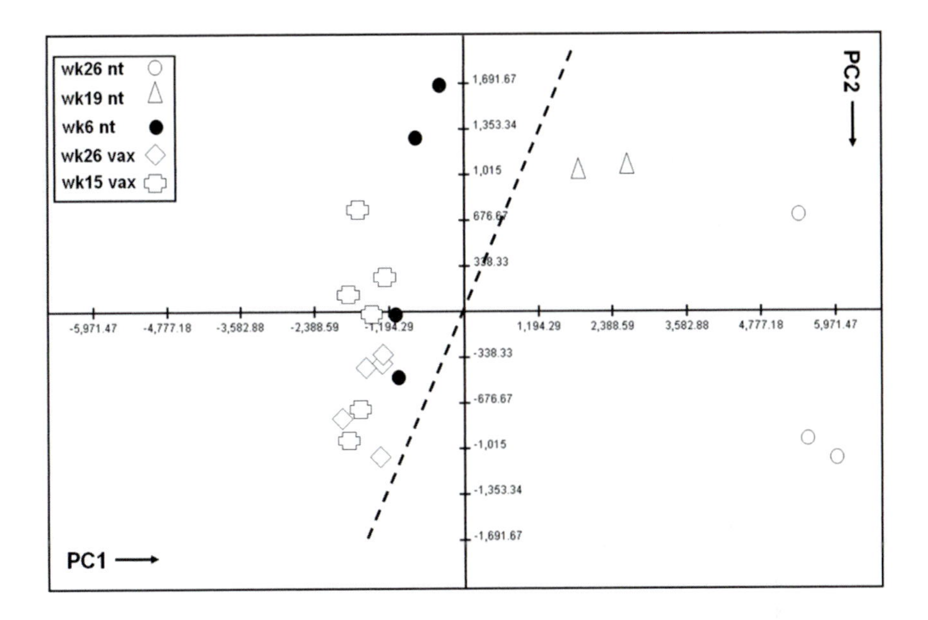

Figure 7.1 PCA analysis of global gene expression changes in not-treated animals at weeks 6, 19 and 26 (wk6nt, wk19nt, wk26nt) and vaccinated animals at weeks 15 and 26 (wk15vax and wk26vax). The vaccination effect divides the microarray data into two groups. The vax profiles cluster with wk6nt and are separate from the other two nt profiles.

between the samples. The vaccination effect allows separation of the microarray data into two groups (Fig. 7.1, dashed line). PCA shows that tumors in wk26nt are different from those of wk19nt. These differences at the transcriptional profile level are linked to both tumor progression and change in the number of tumor cells in the mammary glands. The transcription profiles generated from wk15vax or wk26vax mice are very similar to those at the time when vaccination began (week 6) and overall differences between samples are more limited, since they are solely associated with the second component and this accounts for only 9% of the total variance (Fig. 7.1). These results indicate that tumor progression is stably halted by Triplex vaccination (Astolfi *et al.*, 2005a).

Early changes in gene expression induced by vaccination were investigated by comparing the expression patterns of mammary tissues from wk15nt and wk15vax. A total of 155 probe sets were differentially expressed (Fig. 7.2A). Genes more expressed in vaccinated samples were almost exclusively immune response genes (Fig. 7.2A, red bar). Ig chains, IFN-γ-induced genes and, to a lesser extent, inflammation markers are those highly induced in vaccinated with respect to non-treated samples. This suggests that the predominant immunological response is based on B-cells and IFN-γ-secreting T-cells. Other immune cell populations such as neutrophils and dendritic cells play a minor role, as exemplified by the lower induction of their specific genes, and the same is true of cytotoxic T-cells (Astolfi *et al.* 2005a).

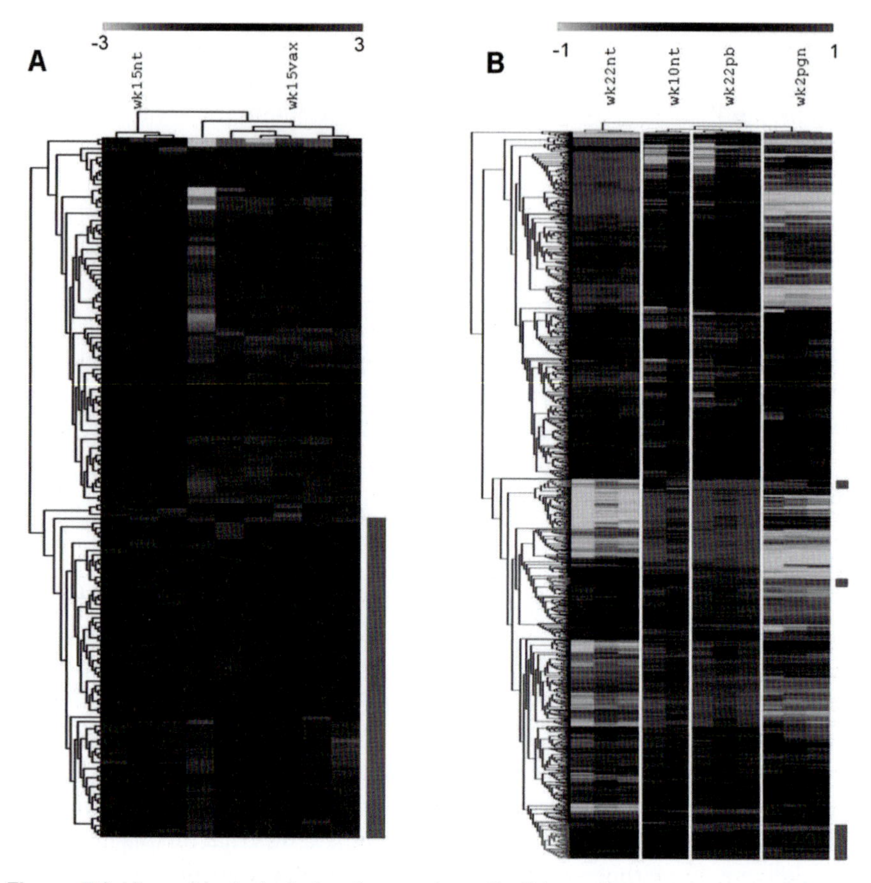

Figure 7.2 Hierarchical clustering (parameters: Euclidean distance, average linkage clustering) of genes differentially expressed by Astolfi *et al.* (2005a) and Quaglino *et al.* (2004). Data are represented as the variation of each sample with respect to the mean of all experimental points. Clusters (red bars) represent genes mainly over-expressed in vaccinated samples and tightly related to immune response functions. Clustering was performed: (A) on genes differentially expressed at 15 weeks between not-treated and Triplex vaccinated mice; B) on wk2pgn, wk10nt, wk22nt, wk22pb. Color figure available at http://bioinformatica.unito.it/bioinformatics/CavalloBook2006/Fig7.2.pdf

Prime-and-boost vaccination

In this case, the mechanisms involved were investigated by comparing the gene expression patterns of the mammary glands of wk10nt and wk22nt mice with those of 22-week-old primed-and-boosted (wk22pb) mice. Those displayed by pregnant mice (wk2pgn) were included to represent physiological hyperplasia. Total RNA extracted individually from the mammary tissue was pooled (three animals per pool) to obtain biological triplicates for wk2pgn, wk22nt and wk22pb, and a biological duplicate for wk10nt. The pools were used to synthesize biotinylated cRNAs and hybridized on 11 MG-U74Av2 Affymetrix GeneChips. To identify genes differentially expressed, a funnel-shaped procedure followed by statistical analysis applied to the microarray data yielded 1968 genes associated with 2179 probe sets (Quaglino *et al.*, 2004). PCA analysis of wk10nt, wk22nt, and wk22pb glands showed that the dataset is mainly distributed on the first two principal components, and that each experimental group is characterized by good reproducibility as all replicates cluster together (Calogero *et al.* 2004). The clustering data, as in the case of the Triplex vaccination, indicate that the transcription profile of wk22pb is remarkably different from that of wk22nt and is very similar to that of wk10nt (Fig. 7.2B). This observation has strong biological implications, as it perfectly matches the *in vivo* observations and indicates that carcinogenesis is arrested at the atypical hyperplasia stage in primed and boosted mice (Quaglino *et al* 2004). Nevertheless, despite this great homology of gene expression profiles, a limited number of probe sets (Fig. 7.2B, red bars) were differentially expressed in wk22pb with respect to wk10nt. All these probe sets are upmodulated (> 1-fold) in wk22pb with respect to wk10nt and related to immune-response (Quaglino *et al.*, 2004). Most are linked to genes generically pertinent to the humoral response and encoding antibody-related genes. The gene coding for the Ig J polypeptide is one of those selectively upregulated and suggests local production of IgA (Quaglino *et al.*, 2004). This indication fits in well with the observation of numerous plasma cells in the mesenchyma and in the stroma surrounding the "frozen" lesions (Lo Iacono *et al.*, 2005). In addition, gene expression profiles in wk10nt and wk22pb show that genes related to the HER2 signal transduction pathways, such as ras, cyclin D1, cdk4 cyclin-dependent kinase, jun transcription factor, protein kinase C and PI3K are downregulated compared to wk22nt. By contrast, transcription of the Cbl, endophilin, and ubiquitin genes related to the HER2 degradation pathway is upregulated. The inhibited HER2 transduction pathway results in reduction of proliferation (Quaglino *et al.*, 2004). This is a further confirmation of the role of the humoral immune response. *In vitro* studies with confocal microscopy, in fact, have demonstrated that anti-HER2 antibodies in the sera of vaccinated mice downregulate HER2 from the membrane of tumor cells that over-express it (Fig. 7.3).

Deriving tumor-associated antigens (TAAs) from BALB-neuT breast carcinoma transcription profiles

How far can the results of these two studies be transferred to the clinical setting? Could HER2-based vaccination halt the progression of human *in situ* carcinoma by causing a "Herceptin-like effect," since most patients responding to Herceptin develop resistance within a year (Vogel *et al.*, 2002)? Vaccination with a combination of HER2 and with other TAAs might be more effective? Like HER2, an ideal TAA should be necessary to maintain the malignancy of the neoplastic cells highly expressed at all tumor stages, but not in normal tissue. Characterization of a distinct set of differentially expressed genes/proteins by cDNA technology or proteomics might contribute to the identification of targets and result in the design of novel treatment strategies. For this purpose

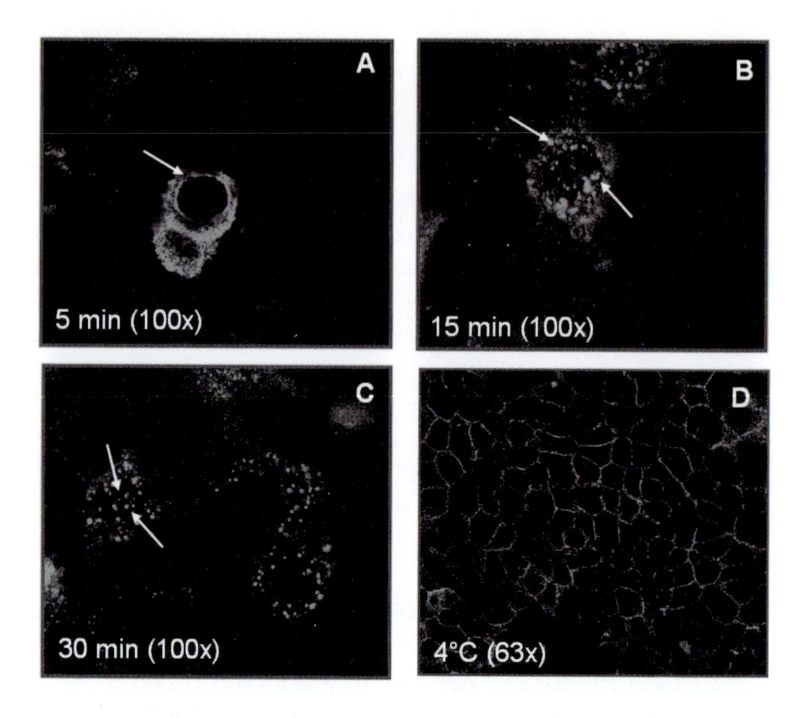

Figure 7.3 Vaccination-induced antibodies downregulate HER2 membrane expression in HER2⁺ derived from a BALB-neuT mammary tumor (TUBO cell line) and incubated with the immune serum from wk22pb mice at 37°C (A–C) and 4°C (D). Cytoplasmic dots showing HER2 internalization after different incubation times are indicated by arrows.

a meta-analysis was performed of all the microarray data from the two studies. This combination was possible since it has been demonstrated that if microarray analyses are carried out accurately, the results produced (using the same animal model and technological platform) by laboratories with slightly different designs, but addressing the same biological question, are strikingly comparable (Astolfi *et al.*, 2005b).

The evidence that the number of neoplastic cells in the mammary tissue of BALB-neuT mice increases constantly during aging (Astolfi *et al.*, 2005b; Cavallo *et al.*, 2005) was used to select only genes already expressed in the mammary tissues from 6-week-old animals and whose expression was maintained in older animals. 204 genes were identified by fitting a linear model to the transcription profiles at wk6nt, wk10nt, wk19nt, wk22nt and wk26nt (Cavallo *et al.*, 2005). The identified genes were characterized by significant enrichment in gene ontology (Harris, *et al.* 2004) classes associated with the extracellular compartment (Table 7.1).

To fulfill the ideal TAA requirements, only five of these genes were regarded as potentially interesting (Tes, Rcn2, Rnf4, Cradd, Galnt3). Their expression level in both normal (Yanai *et al.*, 2005) and tumor tissues (van 't Veer *et al.*, 2002) was nearly comparable to that of HER2 (Cavallo *et al.*, 2005).

Table 7.1 GO cellular components (CC) and biological process (BP) classes within the 204 genes found linearly linked to tumor mass increase.

GO class	Description	*P*-value	GO
GO:0007155	Cell adhesion	$< 10^{-3}$	BP
GO:0030198	Extracellular matrix organization and biogenesis	5×10^{-3}	BP
GO:0007160	Cell-matrix adhesion	5×10^{-3}	BP
GO:0016327	Apicolateral plasma membrane	4×10^{-3}	CC
GO:0016020	Membrane	5×10^{-3}	CC
GO:0030054	Cell junction	6×10^{-3}	CC
GO:0005923	Tight junction	7×10^{-3}	CC
GO:0005911	Intercellular junction	7×10^{-3}	CC
GO:0030057	Desmosome	8×10^{-3}	CC
GO:0005578	Extracellular matrix	8×10^{-3}	CC

Tes maps to a fragile site on chromosome 7q31.2 (Han *et al.*, 2003) and may have a tumor suppression activity (Sarti *et al.* 2005). Rcn2, reticulocalbin, has an unknown function and it is implicated in tumor cell invasiveness (Liu *et al.*, 1997). Rnf4 acts as a transcription regulator (Kaiser *et al.*, 2003). When ectopically expressed, it inhibits the proliferation of both somatic and germ cell tumor-derived cells (Pero *et al.* 2001; Hirvonen-Suntti *et al.*, 2004). Cradd encodes a death domain (CARD/DD)-containing protein and induces apoptosis (Ahmad *et al.*, 1997; Duan *et al.*, 1997). Galnt3 encodes UDP-GalNAc transferase 3, a member of the GalNAc-transferases family. Its expression is associated with the differentiation and aggressiveness of ductal adenocarcinoma of the pancreas (Yamamoto *et al.*, 2004).

Bioinformatic approaches for microarray data analysis

Computational tools are essential in microarray data analysis and mining to extract information knowledge from the experimental results. In this section we present some of the approaches used in transcriptional profiling studies of HER2 transgenic mice. All approaches described in this section are focused on Affymetrix GeneChip (Affymetrix, Santa Clara, CA) data analysis, since all the transcriptional profiling experiments were done with this technological platform (Quaglino *et al.*, 2004; Astolfi *et al.*, 2005b).

Oligonucleotide chips

In GeneChips a probe set composed of 11–20 25mer oligonucleotides (probes) is used to monitor a transcript or EST. Probes belonging to the same set are distributed to different locations on the chip. Each probe, also called perfect match (PM), is associated with a "negative control" oligonucleotide, called mismatch (MM), whose sequence is equal to that of PM apart from a single central nucleotide mismatch that strongly destabilizes hybridization of the target molecule: PM/MM is called the probe pair. The definition of the probe set intensity summary might be calculated, depending on the chosen algorithm, using only the PM or the complete probe pair, with MM being used as the measure of nonspecific hybridization (Irizarry *et al.*, 2003a, b).

GeneChips data analysis

The main steps in a microarray data analysis are:

- identification of possible hybridization or production artifacts and analysis of probe intensity distribution;
- replicates quality evaluation;
- probe set intensity summary;
- data normalization;
- data filtering;
- statistical validation;
- data annotation.

All GeneChip data manipulations described here were performed with Bioconductor packages (Gentleman *et al.* 2004). An entry-level tutorial on Bioconductor is available at http://www.bioinformatica.unito.it/bioinformatics/embo.course/milano2005.html.

Identification of possible hybridization or production artifacts and analysis of probe intensity distribution

Hybridization/washing artifacts can be detected using the "fitPLM" function implemented in the "affyPLM" Bioconductor package. This function produces pseudo-images (Fig. 7.4A and B) which are very useful for detecting artifacts on arrays. Local artifacts (Fig. 7.4A) affecting less than 10% of the overall array surface have limited impact on probe set intensity calculation, since probe pairs of the same probe set are scattered all over the array. Instead, if artifacts affect the vast majority of an array, the array should be discarded (Fig. 7.4B).

Another quality control parameter is the overall distribution of probe intensities. This is evaluated using the "hist" function implemented in the "affy" Bioconductor package (Gautier *et al.*, 2004). The distribution of probe intensities should have a log-normal shape and be spread over more than two log units (Fig. 7.4C). A too narrow distribution frequently indicates that the amount of cRNA used in the experiment was too small (Fig. 7.4D) and many PM signals will result below the instrument sensitivity.

Replicate quality evaluation

An important issue in microarray analysis is the quality of replicates. To perform statistical validation analysis, each experimental point should be replicated a few

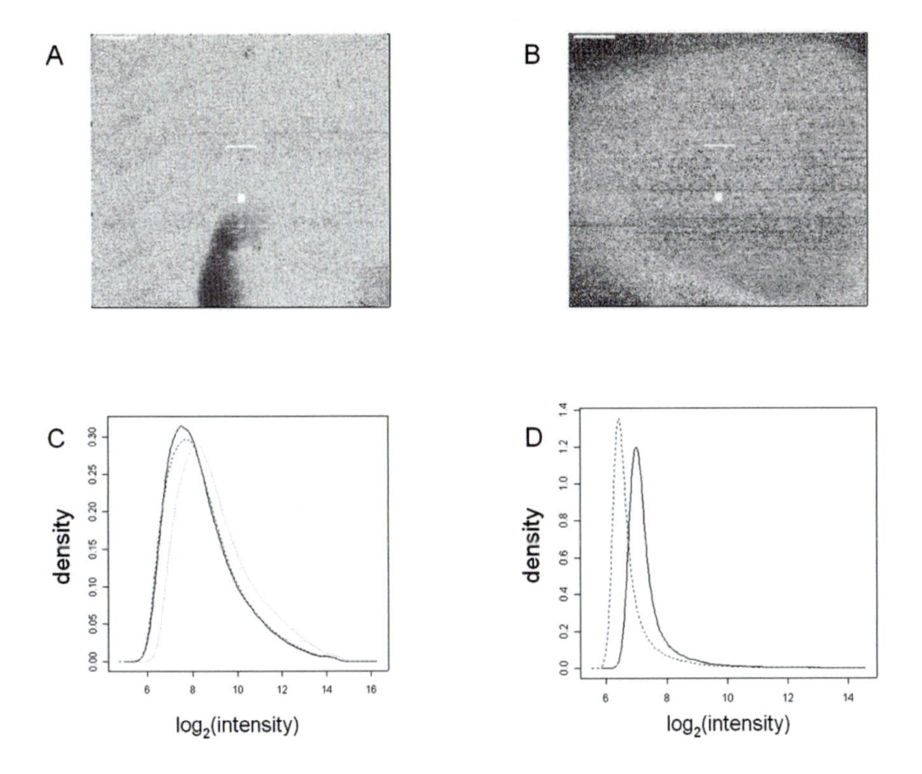

Figure 7.4 Pseudo-images generated by the fitPLM function (affyPLM Bioconductor package). Positive residuals are represented in red and negative residuals in blue. Strong colors represent strong deviation from the calculated probe set model. (A) Artifact with limited effect on the overall quality of the array. (B) Artifact affecting the overall performance of the array. PMs density distribution plots: (C) PM distribution is log-normal over a large interval of log2(intensities). (D) The amount of cRNA used in the hybridization was over-estimated and the overall distribution of the PM signals is very narrow. Color figure available at http://bioinformatica.unito.it/bioinformatics/CavalloBook2006/Fig7.4.pdf

times to estimate the technical and experimental variability associated with the experiment. Replicates are of two types: technical and biological. While recent improvement in GeneChips technology has made technical replicates inessential, biological replicates are still essential, because the main source of variation is related to the biological issue under study. In the case of animal studies, they are provided by collecting total RNA from different animals of the same treatment group. The minimal number of replicates needed to obtain a good estimation of variance can be predicted in different ways (Altman, 2005; Pan *et al.*, 2002; Wei *et al.*, 2004; Wang *et al.*, 2004), for example 8–10 replicates for experiments based on inbred animals. However, due to cost and/or limitation in the amount of biological material available, three or four replicates are considered sufficient if sufficient homogeneous. The *r*-squared coefficient, the fraction of variance explained by a linear model, is the parameter most frequently used to evaluate homogeneity. 1 indicates that the two sets of data are identical, while 0 indicates the absence of similarity between samples. Good biological replicates are usually characterized by an *r*-squared ≥ 0.8. However, this estimation of homogeneity does not consider the presence of a strong probe effect (Li *et al.*, 2001a,b). This a peculiarity of GeneChips and is represented by the differences between signals associated with different probes of the same probe set, which are more substantial than those between the two spike-in concentrations at single probe level (Fig. 7.5A). A more efficient way to evaluate homogeneity has been introduced by Irizarry (Irizarry *et al.* 2005) and is based on the use of a reference array against which biological replicates are compared. The homogeneity within replicates is then evaluated at the level of absolute fold change variation of each probe set with respect to the reference. Each biological replicate is described as the sorted list (fold change variation in descending order) of all probe sets. Homogeneous replicates must have a concordance in probe set composition at the top rank positions (e.g., 100–1,000) of at least 40%, corresponding to a *r*-squared between replicates greater than 0.8 (Irizarry *et al.*, 2005). A graphical representation (CAT plot) of this concordance analysis is illustrated in Fig. 7.5B.

Probe set intensity summary

Many algorithms have been described to overcome the complex problems related to probe set intensity calculation: MAS (www.affymetrix.com), DCHIP (Li *et al.* 2001a,b), RMA (Irizarry *et al.*, 2003a), GCRMA (http://www.biostat.jhsph.edu/~ririzarr/papers/p177-irizarry.pdf). The probe set summaries for

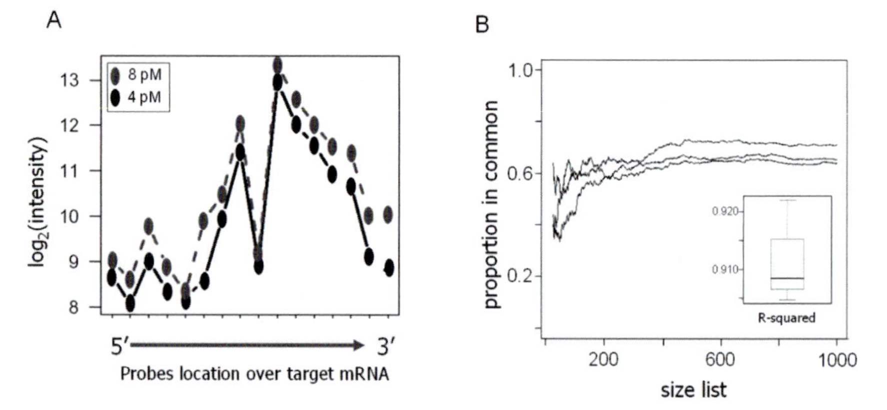

Figure 7.5 Replicate quality analysis. (A) Intensities of all probes associated with a given probe set when the target cRNA is hybridized at 4 and 8 picomolar spike-in concentrations. B) CAT plot showing curves of concordance between three replicates versus a reference sample, for lists ranging from 25 to 1000. The concordance is good since the proportion in common is at least 0.6 for lists greater than 200. The inset box shows the quality of replicates as the distribution of r-squared values within all replicates. These values are good (> 0.9).

the transcription profiles of wk10nt, wk22nt and wk22pb described above were calculated both with RMA (Quaglino *et al.*, 2004) and GCRMA (Lo Iacono *et al.*, 2005). GCRMA represents the best compromise between the signal precision measurement of the original MAS algorithm, developed by Affymetrix, and the signal accuracy measurement of the more recent RMA approach.

Data normalization

Data normalization guarantees that observed differences in intensities are really due to differential expression and not to technical artifacts. Affymetrix has proposed that intensities should be scaled such that each array has the same average value (constant normalization, in affy Bioconductor package). This approach, however, is less effective in the case of non-linear relationships between arrays, which are efficiently dealt with by the approaches of Li and Wong (2001b) and Bolstad (2003).

Li and Wong (2001b) have proposed an "invariant set normalization," a procedure based on non-differentially expressed genes. Since these are generally not known *a priori*, they are determined based on intensity ranks that are similar between two arrays. An iterative procedure identifies this invariant set by choosing probes with a small enough proportional rank difference (absolute rank difference divided by the total number) where small enough depends on the average intensity rank in the two arrays. A piecewise-linear running median curve is then calculated and used as the normalization curve.

Bolstad proposes two methods. Both are implemented in the affy Bioconductor package:

- *Cyclic loess*. This approach stems from a M versus A plot, where M is the difference in log expression values and A is their average (Dudoit *et al.*, 2002). A normalization curve is fitted to this M versus A plot by using loess, which is a method of local regression (Cleveland and Devlin, 1988). The fit based on the normalization curve is subsequently subtracted from the M values. However, rather than being applied to two-color channels on the same array, as is done for the cDNA case, normalization is applied to probe intensities from two arrays. To deal with more than two arrays, the method is extended to look at all distinct array pairwise combinations.
- *Quantile normalization*. The goal of the quantile method is to make the distribution of probe intensities equal across arrays. If the distribution of two data vectors is the same a quantile-quantile plot produces a straight diagonal line. This concept can be extended to n dimensions by organizing the arrays to be normalized in a matrix (columns represent arrays, rows PMs). Columns are then sorted in descending order, averages across rows are calculated and assigned to each element of the rows. Columns are then reordered as the original matrix to generate the normalized set.

We have employed a benchmark experiment (Choe *et al.*, 2005) to determine which method provides the best normalization by determining how normalization affects the identification of "spike-in" differentially expressed genes that cannot be detected as average fold changes (fc) in the raw data. The Choe (2005) dataset is ideal for this analysis. because it contains 1,309 individual cRNAs with known relative concentrations between the spike-in and control samples. We selected 325 probe sets with an fc greater than 2 (188 sets) or equal to 2 (137 sets), and used equation (1) to assess the effect of each normalization method/approach on average fc evaluation.

$$enrichment = \frac{\left|N_{fcnormalized}\right| \geq 2}{\left|N_{fcraw}\right| \geq 2} \times 100 \qquad (7.1)$$

where $|N_{fcnormalized}| \geq 2$ represents the number of probe sets, which are characterized by an average fc ≥ 2 after normalization, and $|N_{fcrawd}| \geq 2]$ represents the number of probe sets characterized by an average fc ≥ 2 in the raw dataset.

A normalization approach that improves overall data quality should have an "enrichment" value greater than that given by raw data only (i.e., 100). If the fc is greater than two invariant sets, quantile and cyclic-loess methods perform comparable and better than the constant method (Fig. 7.6A). However, if fc is equal to 2, cyclic-loess normalization performs better than the other methods (Fig. 7.6B).

Data filtering

The premise of this important step in microarray data analysis is removal of genes deemed to be not expressed according to some specific criterion under the control of the user (von Heydebreck, *et al* 2004). It can also be used to eliminate genes that do not show sufficient variation in expression across all samples and therefore have little discriminatory power. There are various approaches to data filtering:

- removing probe sets with an intensity below a user-defined threshold (von Heydebreck *et al.*, 2004);
- removing probe sets classed as "absent" in the expression call algorithm developed by Affymetrix (www.affymetrix.com);
- removing probe sets that do not vary in function of the experimental conditions (von Heydebreck *et al.*, 2004);
- removing probe sets that do not belong to a specific functional class (e.g., transcription factors).

The efficacy of some of these approaches is illustrated at www.bioinformatica.unito.it/bioinformatics/DAII.

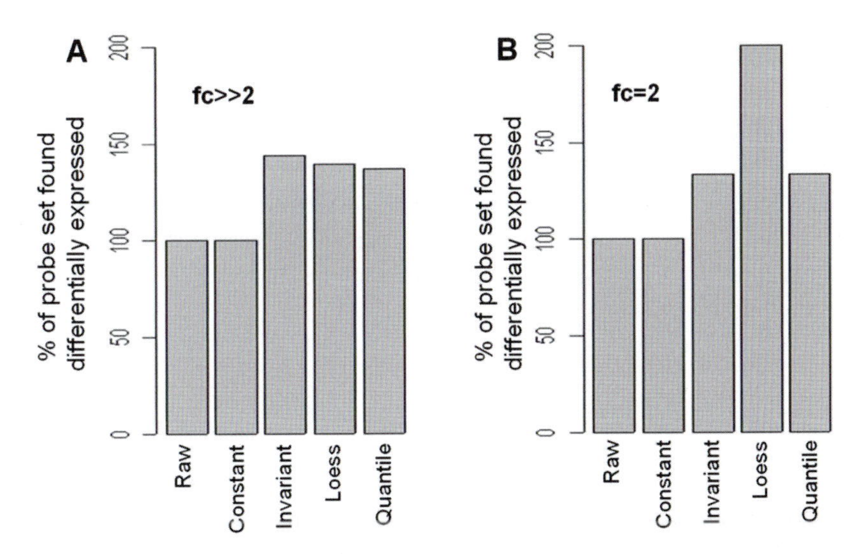

Figure 7.6 Normalization method performances. The yield of true differentially expressed probe sets was investigated in the spike-in experiments from Choe (2005) by comparing the number of such sets identified upon normalization with respect to those identified in the raw data. A) For spike-in with absolute fc > 2, the Li, and Bolstad approaches perform in much the same way and increase the number of true differentially expressed probe sets that can be detected with respect to the raw data and constant normalization according to Affymetrix. (B) Where the absolute fc = 2, the cyclic-loess method gives the best yield.

In the transcriptional profile study by Quaglino (2004), filtering was performed with the funnel-shaped procedure described by Saviozzi *et al.* (2004). All probe sets "absent" in more than 90% of the experiments were selected. Their intensity distribution was calculated and an intensity threshold was defined as the 90th percentile of the distribution. All probe sets with an average intensity smaller than the threshold in all experiments were discarded as "not expressed." The method known as interquantile filter (von Heydebreck *et al.* 2004) eliminates probe sets with insufficient variation in expression across all samples; it was used in the papers by Astolfi (2005a), Lo Iacono (2005) and Cavallo *et al.* (2005)

Statistical validation

Statistical validation is an important issue when identifying differentially expressed genes. Since experiments are generally conducted on a very limited number of replicates and involve many parallel tests (genes to be evaluated for differential expression) the exact opposite of an ideal condition for a statistical test (many replicates, few parallel tests) is present.

A variety of statistical tests. (e.g., *t*-test, ANOVA, linear model analysis, etc.) have been modified and optimized to obtain an estimate of inter/intra sample variance despite the limited number of replicates. In all these cases the underlying assumption is that the expression level of one transcript is independent from all others.

An optimal method of identifying differentially expressed genes cannot be determined. Each method will catch only part of the biologically meaningful differences (Hosack *et al.*, 2003) and results from different methods will in general include a common set of robustly differentially expressed genes. To reduce errors arising from multiple testing (for example Type I error or false positives), the Bioconductor package implements multiple testing procedures (MTP) for controlling a broad class of type I error rates (Dudoit, 2004; van der Laan *et al.*, 2004a,b; Pollard and van der Laan, 2004). A key feature is the generation of a test statistics null distribution that is used to determine thresholds and adjustments of calculated *P*-values.

Choe (2005) has compared the performances of three statistical approaches to the identification of differentially expressed genes: "significance analysis for microarrays" (SAM) (Tusher, 2001), CyberT (Baldi and Long, 2001) and the Student's *t*-statistic. CyberT, which outperformed the other approaches, uses a signal-intensity-dependent standard deviation (SD) and probe sets at low signal intensities obtain reduced significance even when the fold-change value is high. For SAM the *t*-statistic depends more on the fc value than the noise level. The basic *t*-statistic, on the other hand, was affected by false-positives resulting from the artificially low SD due to the limited number of replicates.

In transcriptional profiling studies on HER2 transgenic mice interesting differentially regulated genes were selected using a moderated-t test and ANOVA (Tusher *et al.*, 2001; Baldi and Long, 2001) or by linear model analysis (Smyth, 2004). In the paper by Quaglino (2004), the critical choice of significance thresholds (i.e., p-value, false discovery rate) was investigated relying on intrinsic biological information present in the experiment. Hosack (2003) showed that different combinations of intensity calculation, data normalization and statistical validations produce only partially overlapping sets of differentially expressed genes, even if the highest-ranking gene ontology (GO) terms identified in these subsets were conserved. We therefore used over-represented GO terms (Hosack *et al.*, 2003) to select the optimal significance. A SAM multiclass test (Tusher *et al.*, 2001) was performed at several significance thresholds to assess enrichment in GO classes related to an immune response the main biological event expected to be associated with prime and boost vaccination. A more detailed analysis (http://www.bioinformatica.unito.it/bioinformatics/Forni/additional_info/statistical_validation_approach.html) shows that there is a specific threshold condition that reveals the highest number of statistically enriched GO classes linked to immunological events, combined with the highest percentage of genes within those classes. Using an arbitrary threshold definition, a number of these genes might instead have not been detected. Use of biological information when defining the optimal statistical threshold is therefore a more efficient approach than the use of an arbitrary definition simply based on statistical parameters (e.g., false discovery rate).

Data annotation

Another important issue in microarray data analysis is the specific association of probe set identifiers with genome annotated transcripts. GeneChips probes are designed starting from clusters in the most recent Unigene release (www.ncbi.nlm.nih.gov/entrez/query.fcgi?db = unigene) trying, at the same time, to reduce unwanted cross-hybridization with other clusters. Probes are however not updated at every Unigene rebuild, but only when a new GeneChip version becomes available. After a few Unigene releases, therefore, a probe set that initially had all probes mapped to a specific transcript now may map only a subset to the target sequence reducing the specificity of target recognition (Mecham *et al.*, 2004a,b). On the other hand, addition of new sequence clusters to Unigene can reveal possible cross-hybridization to other transcripts. All this information is available through the quarterly updated Affymetyrix on-line annotation tool, Netaffx (www.affymetrix.com/analysis/netaffx/index.affx). Three additional public tools (AnnBuilder at www.bioconductor.org/packages/data/annotation/stable/src/contrib/html/;RESOURCERER at www.tigr.org/tigr-scripts/magic/r1.pl; Ensembl at www.ensembl.org/Multi/martview) are also available using different approaches to associate a probe set with a specific transcript.

In our opinion, the best association between probe set and transcript is the one available at the Affymetrix site, since it is based on direct mapping of probes on transcript sequences. In addition much information on probe specificity and cross-mapping is available and can be used by experienced microarray users to obtain a clear estimate of probe set quality.

Conclusions

Microarray data that extensively monitor gene expression in human tumors and in their murine models counterpart will offer an unprecedented opportunity to identify and test tumor antigens that can be targeted by specific immunopreventive approaches, whose efficacy and safety can easily be evaluated in the murine model.

From the studies described in this chapter it is clear that genes and the immune response patterns shown by microarray analysis perfectly mirror the results of more conventional immunological, morphological and immunohistochemical studies. Further support is thus provided for the view that in this immunopreventive protocol antibodies and IFN-γ are the dominant immune responses elicited by vaccination. From the two studies (Quaglino *et al.*, 2004; Astolfi *et al.*, 2005a) it is unmistakable that the vaccination "freeze" mammary transcriptome at a preneoplastic stage and blocked the expression of many genes related to the development of mammary carcinoma. At the same time, vaccinations halt the progression of a tumor at the stage it is in when they begin.

Our data also indicate that if the experimental design and the biological part of the microarray analysis are carried out accurately, the results produced by different laboratories, but addressing the same biological question, are strikingly comparable. This general observation shows that microarray analysis is now mature and ensures a good level of inter-laboratory reproducibility.

Frequently raised questions are:

- Can microarrays really yield any worthwhile new and valuable information?
- Do transcriptional profiling studies corroborate what can be found by integrated pathological and immunological analysis?

We believe that the prime advantage of the microarray technology is related to the results provided, in addition to confirming those offered by more conventional techniques, which will in most cases be unexpected or difficult to predict. However, unexpected data can be very useful to generate new hypotheses that might open the way to a better comprehension of cancer progression and metastasis and to the design of new applications that can be translated into clinical applications.

Acknowledgments

This work was supported by grants from: the Italian Association for Cancer Research (AIRC); the Italian Ministry for Education, the Universities and Research (MIUR), FIRB projects RBAU012RLC, RBNE017B4C, RBAU01JTHS, RBNE0157EH; the Italian Ministry of Health; the University of Turin. We thank Prof. John Iliffe for his critical review of the manuscript and for helpful discussions.

References

Ahmad, M., Srinivasula, S.M., Wang, L., Talanian, R.V., Litwack, G., Fernandes-Alnemri, T., and Alnemri, E.S. (1997). CRADD, a novel human apoptotic adaptor molecule for caspase-2, and FasL/tumor necrosis factor receptor-interacting protein RIP. Cancer Res. *57*, 615–619.

Altman, N. (2005). Replication, variation and normalisation in microarray experiments. Appl. Bioinformatics *4*, 33–44.

Astolfi, A., Landuzzi, L., Nicoletti, G., De Giovanni, C., Croci, S., Paladini, A., Ferrini, S., Iezzi, M., Musiani, P., Cavallo, F., Forni, G., Nanni, P., and Lollini, P.L. (2005a). Gene expression analysis of immune-mediated arrest of tumorigenesis in a transgenic mouse model of HER-2/neu-positive basal-like mammary carcinoma. Am. J. Pathol. *166*, 1205–1216.

Astolfi, A., Rolla, S., Nanni, P., Quaglino, E., De Giovanni, C., Iezzi, M., Musiani, P., Forni, G., Lollini, P.L., Cavallo, F., and Calogero, R.A. (2005b). Immune prevention of mammary carcinogenesis in HER-2/neu transgenic mice: a microarray scenario. Cancer Immunol. Immunother. *54*, 599–610.

Baldi, P., and Long, A.D. (2001) A Bayesian framework for the analysis of microarray expression data: regularized t -test and statistical inferences of gene changes. Bioinformatics *17*, 509–519.

Boggio, K., Nicoletti, G., Di Carlo, E., Cavallo, F., Landuzzi, L., Melani, C., Giovarelli, M., Rossi, I., Nanni, P., De Giovanni, C., Bouchard, P., Wolf, S., Modesti, A., Musiani, P., Lollini, P.L., Colombo, M.P., and Forni, G. (1998) Interleukin 12-mediated prevention of spontaneous mammary adenocarcinomas in two lines of Her-2/neu transgenic mice. J. Exp. Med. *188*, 589–596.

Bolstad, B.M., Irizarry, R.A., Astrand, M., and Speed, T.P. (2003). A comparison of normalization methods for high density oligonucleotide array data based on variance and bias. Bioinformatics *19*, 185–193.

Calogero, R.A., Musiani, P., Forni, G., and Cavallo, F. (2004). Towards a long-lasting immune prevention of HER2 mammary carcinomas: directions from transgenic mice. Cell Cycle 3, 704–706.

Cavallo, F., Astolfi, A., Iezz,i M., Cordero, F., Lollini, P.L., Forni, G., and Calogero, R. (2005). An integrated approach of immunogenomics and bioinformatics to identify new Tumor Associated Antigens (TAA) for mammary cancer immunological prevention. BMC Bioinformatics 6, Suppl 4:S7.

Choe, E.S., Boutros, M., Michelson, A.M., Church, G.M., and Halfon, M.S. (2005). Preferred analysis methods for Affymetrix GeneChips revealed by a wholly defined control dataset. Genome Biol. 6, R16.

Cleveland, W.S. Devlin, S.J. (1998). Locally-weighted regression: an approach to regression analysis by local fitting. J. Am. Stat. Assoc., *83*, 596–610.

De Giovanni, C., Nicoletti, G., Landuzzi, L., Astolfi, A., Croci, S., Comes, A., Ferrini, S., Meazza, R., Iezzi, M., Di Carlo, E., Musiani, P., Cavallo, F., Nanni, P., and Lollini, P.L. (2004). Immunoprevention of HER-2/neu transgenic mammary carcinoma through an interleukin 12-engineered allogeneic cell vaccine. Cancer Res. *64*, 4001–4009.

Di Carlo, E., Diodoro, M.G., Boggio, K., Modesti, A., Modesti, M., Nanni, P., Forni, G., Musiani, P. (1999). Analysis of mammary carcinoma onset and progression in HER-2/neu oncogene transgenic mice reveals a lobular origin. Lab. Invest. *79*, 1261–1269.

Duan, H., Dixit, V.M. (1997). RAIDD is a new 'death' adaptor molecule. Nature *385*, 86–89.

Dudoit, S., Yang, Y.H., Callow, M.J., Speed, T.P. (2002). Statistical methods for identifying genes with differential expression in replicated cDNA microarray experiments. Stat. Sin. 12, 111–139.

Dudoit, S., van der Laan, M.J., and Pollard, K.S. (2004). Multiple testing. Part I. Single-step procedures for control of general Type I error rates. Statistical Applications in Genetics and Molecular Biology, 3, Article 13. URL http://www.bepress.com/sagmb/vol3/iss1/art13/.

Gautier, L., Cope, L., Bolstad, B.M., and Irizarry, R.A. (2004). affy—analysis of Affymetrix GeneChip data at the probe level. Bioinformatics *20*, 307–315.

Gentleman, R.C., Carey, V.J., Bates, D.M., Bolstad, B., Dettling, M., Dudoit, S., Ellis, B., Gautier, L., Ge, Y., Gentry, J., Hornik, K., Hothorn, T., Huber, W., Iacus, S., Irizarry, R., Leisch, F., Li C., Maechler, M., Rossini, A.J., Sawitzki, G., Smith, C., Smyth, G., Tierney, L., Yang, J.Y., and Zhang, J. (2004). Bioconductor: open software development for computational biology and bioinformatics. Genome Biol. 5, R80.

Han, S.Y., Druck, T., Huebner, K. (2003). Candidate tumor suppressor genes at FRA7G are coamplified with MET and do not suppress malignancy in a gastric cancer. Genomics *81*, 105–107.

Harris, M.A., Clark, J., Ireland, A., Lomax, J., Ashburner, M., Foulger, R., Eilbeck, K., Lewis, S., Marshall, B., Mungall, C., Richter, J., Rubin, G.M., Blake, J.A., Bult, C.,

Dolan, M., Drabkin, H., Eppig, J.T., Hill, D.P., Ni, L., Ringwald, M., Balakrishnan, R., Cherry, J.M., Christie, K.R., Costanzo, M.C.., Dwight, S.S., Engel, S., Fisk, D.G., Hirschman, J.E., Hong, E.L., Nash, R.S., Sethuraman, A., Theesfeld, C.L., Botstein, D., Dolinski, K., Feierbach, B., Berardini, T., Mundodi, S., Rhee, S.Y., Apweiler, R., Barrell, D., Camon, E., Dimmer, E., Lee, V., Chisholm, R., Gaudet, P., Kibbe, W., Kishore, R., Schwarz, E.M., Sternberg, P., Gwinn, M., Hannick, L., Wortman, J., Berriman, M., Wood, V., de la Cruz, N., Tonellato, P., Jaiswal, P., Seigfried, T., and White, R. (2004). Gene Ontology Consortium. The Gene Ontology (GO) database and informatics resource. Nucleic Acids Res. 32, D258–261.

Hirvonen-Santti, S.J., Sriraman, V., Anttonen, M., Savolainen, S., Palvimo, J.J., Heikinheimo, M., Richards, J.S., Janne, O.A. (2004) Small nuclear RING finger protein expression during gonad development: regulation by gonadotropins and estrogen in the postnatal ovary. Endocrinol. 145, 2433–2444.ù

Hosack, D.A., Dennis, G.Jr., Sherman, B.T., Lane, H.C., and Lempicki, R.A. (2003). Identifying biological themes within lists of genes with EASE. Genome Biol. 4, R60.

Irizarry, R.A., Hobbs, B., Collin, F., Beazer-Barclay, Y.D., Antonellis, K.J., Scherf, U., Speed, T.P. (2003a). Exploration, normalization, and summaries of high density oligonucleotide array probe level data. Biostatistics. 4, 249–264.

Irizarry, R.A., Bolstad, B.M., Collin, F., Cope, L.M., Hobbs, B., Speed, T.P. (2003b). Summaries of Affymetrix GeneChip probe level data. Nucleic Acids Res. 31, e15.

Irizarry, R.A., Warren, D., Spencer, F., Kim, I.F., Biswal S., Frank, B.C., Gabrielson, E, Garcia, J.G., Geoghegan, J., Germino, G., Griffin, C., Hilmer, S.C., Hoffman, E., Jedlicka, A.E., Kawasaki, E., Martinez-Murillo, F., Morsberger, L., Lee, H., Petersen, D., Quackenbush, J., Scott, A., Wilson, M., Yang, Y., Ye, S.Q., and Yu, W. (2005) Multiple-laboratory comparison of microarray platforms.Nat. Methods 2, 345–350. Erratum in: Nat. Methods (2005) 2, 477.

Yamamoto, S., Nakamori, S., Tsujie, M., Takahashi, Y., Nagano, H., Dono, K., Umeshita, K., Sakon, M., Tomita, Y., Hoshida, Y., Aozasa, K., Kohno, K., and Monden, M. (2004) Expression of uridine diphosphate N-acetyl-alpha-D-galactosamine: polypeptide N-acetylgalactosaminyl transferase 3 in adenocarcinoma of the pancreas. Pathobiology. 71, 12–18.

Yarden, Y., and Sliwkowski, M.X. (2001). Untangling the ErbB signalling network. Nat. Rev. Mol. Cell Biol. 2, 127–137.

Yanai, I., Benjamin, H., Shmoish, M., Chalifa-Caspi, V., Shklar, M., Ophir, R., Bar-Even, A., Horn-Saban, S., Safran, M., Domany, E., Lancet, D., and Shmueli, O. (2005). Genome-wide midrange transcription profiles reveal expression level relationships in human tissue specification. Bioinformatics 21, 650–659.

Kaiser, F.J., Moroy, T., Chang, G.T., Horsthemke, B, Ludecke, H.J. (2003). The RING finger protein RNF4, a co-regulator of transcription, interacts with the TRPS1 transcription factor. J. Biol. Chem. 278, 38780–38785.

Klapper, L.N., Kirschbaum, M.H., Sela, M., and Yarden, Y. (2000). Biochemical and clinical implications of the ErbB/HER signaling network of growth factor receptors. Adv. Cancer Res. 77, 25–79.

Li, C., and Wong, W.H. (2001a). Model-based analysis of oligonucleotide arrays: expression index computation and outlier detection. Proc. Natl. Acad. Sci. USA. 98, 31–36.

Li, C., and Hung Wong, H. (2001b). Model-based analysis of oligonucleotide arrays: model validation, design issues and standard error application. Genome Biol. 2, RESEARCH0032.

Liu, Z., Brattain, M.G., and Appert, H. (10097). Differential display of reticulocalbin in the highly invasive cell line, MDA-MB-435, versus the poorly invasive cell line, MCF-7. Biochem Biophys. Res. Commun. *231*, 283–289.

Lo Iacono, M., Cavallo, F., Quaglino, E., Rolla, S., Iezzi, M., Pupa, S.M., De Giovanni, C., Lollini, P.L., Musiani, P., Forni, G., and Calogero, R.A. (2005). A limited autoimmunity to p185neu elicited by DNA and allogeneic cell vaccine hampers the progression of preneoplastic lesions in HER-2/NEU transgenic mice. Int. J. Immunopathol. Pharmacol. *18*, 351–363.

Lollini, P.L., and Forni, G. (2003) Cancer immunoprevention: tracking down persistent tumor antigens. Trends Immunol. *24*, 62–66.

Lucchini, F., Sacco, M.G., Hu, N., Villa, A., Brown, J., Cesano, L., Mangiarini, L., Rindi, G., Kindl, S., Sessa, F., *et al.* (1992). Early and multifocal tumors in breast, salivary, harderian and epididymal tissues developed in MMTY-Neu transgenic mice. Cancer Lett. *64*, 203–209.

Mecham, B.H., Wetmore, D.Z., Szallasi, Z., Sadovsky, Y., Kohane, I., and Mariani, T.J. (2004a) Increased measurement accuracy for sequence-verified microarray probes. Physiol. Genomics *18*, 308–315.

Mecham, B.H., Klus, G.T., Strovel, J., Augustus, M., Byrne, D., Bozso, P., Wetmore, D.Z., Mariani, T.J., Kohane, I.S., and Szallasi, Z. (2004b). Sequence-matched probes produce increased cross-platform consistency and more reproducible biological results in microarray-based gene expression measurements. Nucleic Acids Res. 32, e74.

Mosesson, Y., and Yarden, Y. (2004). Oncogenic growth factor receptors: implications for signal transduction therapy. Semin. Cancer Biol. 14, 262–270.

Muller, W.J., Sinn, E., Pattengale, P.K., Wallace, R., and Leder P. (1988). Single-step induction of mammary adenocarcinoma in transgenic mice bearing the activated c-neu oncogene. Cell. 54, 105–115.

Nanni, P., Nicoletti, G., De Giovanni, C., Landuzzi, L., Di Carlo, E., Cavallo, F., *et al.* (2001). Combined allogeneic tumor cell vaccination and systemic IL-12 prevents mammary carcinogenesis in HER-2/neu transgenic mice. J. Exp. Med. 194, 1195–1205.

Pan, W., Lin, J., Le, C.T.. (2002). How many replicates of arrays are required to detect gene expression changes in microarray experiments? A mixture model approach. Genome Biol. 3, RESEARCH0022.

Pero, R., Lembo, F., Di Vizio, D., Boccia, A., Chieffi, P., Fedele, M., Pierantoni, G.M., Rossi, P., Iuliano, R., Santoro, M., Viglietto, G., Bruni, C.B., Fusco, A., Chiariotti, L. (2001). RNF4 is a growth inhibitor expressed in germ cells but not in human testicular tumors. Am. J. Pathol. 159, 1225–1230.

Pollard, K.S., and van der Laan, M.J. (2004). Choice of a null distribution in resampling-based multiple testing. J. Stat. Planning Inference *125*, 85–100.

Pupa, S.M., Tagliabue, E., Menard, S., and Anichini, A. (2005). HER-2: A biomarker at the crossroads of breast cancer immunotherapy and molecular medicine. J. Cell Physiol. *205*, 10–18.

Quaglino, E., Rolla, S., Iezzi, M., Spadaio, M., Musiani, P., De Giovanni, C., Lollini, P.L., Lanzardo, S., Forni, G., Sanges, R., Crispi, S., De Luca, P., Calogero, R., and Cavallo, F. (2004). Concordant morphologic and gene expression data show that a vaccine halts HER-2/neu preneoplastic lesions. J. Clin. Invest. 113, 709–717.

Raychaudhuri, S., Stuart, J.M., Altman, R.B. (2000) Principal components analysis to summarize microarray experiments: application to sporulation time series. Pac. Symp. Biocomput., 455–466.

Sarti, M., Sevignani, C., Calin, G.A., Aqeilan, R., Shimizu, M., Pentimalli, F., Picchio, M.C., Godwin, A., Rosenberg, A., Drusco, A., Negrini, M., and Croce, C.M. (2005)

Adenoviral transduction of TESTIN gene into breast and uterine cancer cell lines promotes apoptosis and tumor reduction in vivo. Clin. Cancer Res. 11, 806–813.
Saviozzi, S., Iazzetti, G., Caserta, E., Guffanti, A., Calogero, R.A. (2004). Microarray data analysis and mining. Methods Mol. Med.94, 67–90.
Smyth, G.K. (2004). Linear models and empirical Bayes for assessing differential expression in microarray experiments. Stat. Appl. Genet. Mol. Biol. 3, Article 1.
Spadaio, M., Lanzardo, S., Curcio, C., Forni, G., and Cavallo, F. (2004). Immunological inhibition of carcinogenesis. Cancer Immunol. Immunother. 53, 204–216.
Tsai, J., Sultana, R., Lee, Y., Pertea, G., Karamycheva, S., Antonescu, V., Cho, J., Parvizi, B., Cheung, F., Quackenbush, J. (2001) RESOURCERER: a database for annotating and linking microarray resources within and across species, Genome Biol. 2, SOFTWARE0002.1–0002.4
Tusher, V.G., Tibshirani, R., and Chu, G. (2001). Significance analysis of microarrays applied to the ionizing radiation response. Proc. Natl. Acad. Sci. USA 98, 5116–5121.
van der Laan, M.J., Dudoit, S., and Pollard, K.S. (2004a). Augmentation procedures for control of the generalized family-wise error rate and tail probabilities for the proportion of false positives. Statistical Applications in Genetics and Molecular Biology, 3, Article 15, URL http://www.bepress.com/sagmb/vol3/iss1/art15/.
van der Laan, M.J., Dudoit, S., and Pollard, K.S. (2004b). Multiple testing. Part II. Step-down procedures for control of the family-wise error rate. Statistical Applications in Genetics and Molecular Biology, 3, Article 14, URL http://www.bepress.com/sagmb/vol3/iss1/art14/.
van't Veer, L.J., Dai H., van de Vijver, M.J., He, Y.D., Hart, A.A., Mao, M., Peterse, H.L., van der Kooy, K., Marton, M.J., Witteveen, A.T., Schreiber, G.J., Kerkhoven, R.M., Roberts, C., Linsley, P.S., Bernards, R., and Friend, S.H. (2002). Gene expression profiling predicts clinical outcome of breast cancer. Nature 415, 530–536.
Vogel, C.L., Cobleigh, M.A., Tripathy, D., Gutheil, J.C., Harris, L.N., Fehrenbacher, L., Slamon, D.J., Murphy, M., Novotny, W.F., Burchmore, M., Shak, S., Stewart, S.J., and Press, M. (2002) Efficacy and safety of trastuzumab as a single agent in first-line treatment of HER2-overexpressing metastatic breast cancer. J. Clin. Oncol. 20, 719–726.
von Heydebreck, A., Huber, W., and Gentleman, R. (2004). "Differential Expression with the Bioconductor Project." Bioconductor Project Working Papers. Working Paper 7. http://www.bepress.com/bioconductor/paper7/
Wei, C., Li, J., and Bumgarner, R.E. (2004). Sample size for detecting differentially expressed genes in microarray experiments. BMC Genomics 5, 87–89.
Wang, S.J., and Chen, J.J.. (2004). Sample size for identifying differentially expressed genes in microarray experiments. J. Comput. Biol. *11*, 714–726.
Wheeler, D..L, *et al.* (2003). Database Resources of the National Center for Biotechnology. Nucleic Acids Res. *31*, 28–33.

DNA Microarrays: Beyond mRNA

8

Armin Lahm

Abstract
Application of DNA microarrays in the transcriptional profiling of mRNA has reached practical all areas of modern biomedicine. An extensive literature documents the large variety of biological problems this technique has been applied to, examples of which are described in the previous chapters. Here a number of additional applications will be briefly outlined. From a transcription-centered point of view genome-wide detection of mRNA transcript variants and monitoring of non-coding RNA molecules like microRNAs have gained growing attention and can now be approached with the available technology. On the other hand, variations in the primary genetic information like methylation, nucleotide polymorphisms or amplifications/deletions are increasingly used to explain biological and biomedical phenomena. Moreover, combing several of these complementary views of a biological system has allowed to implement integrative approaches linking a macroscopic phenotype with the underlying genetic background and the complex picture present at the transcriptional level. Proteins have encountered DNA microarrays and ChIP-to-chip experiments are revealing details of chromatin dynamics and the transcriptional state of the genome. As technology proceeds and our understanding of the biological phenomena becomes more refined and detailed, each of these omics-based information sources will contribute to obtain a more integrated view of the complexity of biological systems.

Introduction

Measuring mRNA abundance or differential expression using DNA microarrays has by far been the most widely used application. One area where microarrays have made a major impact on progress is certainly cancer. Expression profiling of many tumor types has allowed us to obtain a detailed molecular view of cancer and to correlate molecular features with clinical parameters.(van't Veer *et al.*, 2002; Hanash, 2004; Segal *et al.*, 2004; Rhodes and Chinnaiyan, 2005; Segal *et al.*, 2005; Bild *et al.*, 2006; Downward, 2006; Whitfield *et al.*, 2006). In addition to canonical transcriptional profiling focusing on mRNA there are however many other ways to utilize DNA microarrays in basic and applied biomedical research (Fig. 8.1). The following paragraphs will provide an overview on major applications together with examples from the literature.

mRNA variants (splicing)

For a long time transcriptional profiling has been focused on monitoring mRNA abundance or differential expression using probes derived from EST sequences or, more general, from sequence information available in major sequence databases (Modrek and Lee, 2002) However, one disadvantage of this approach relies in the fact that only already known variants are monitored making the discovery of new transcript isoforms difficult or even impossible. In particular, variants that share the same 3′ untranslated region and differ only by the presence/absence of one or more internal exons will remain undetected. A more systematic approach taking into account the complete (known or predicted) exon structure of a gene overcomes this deficit and is schematically outlined in Fig. 8.2. Each single exon is monitored separately by exon-specific probes (size of the exon allowing) and boundaries between consecutive exons are monitored

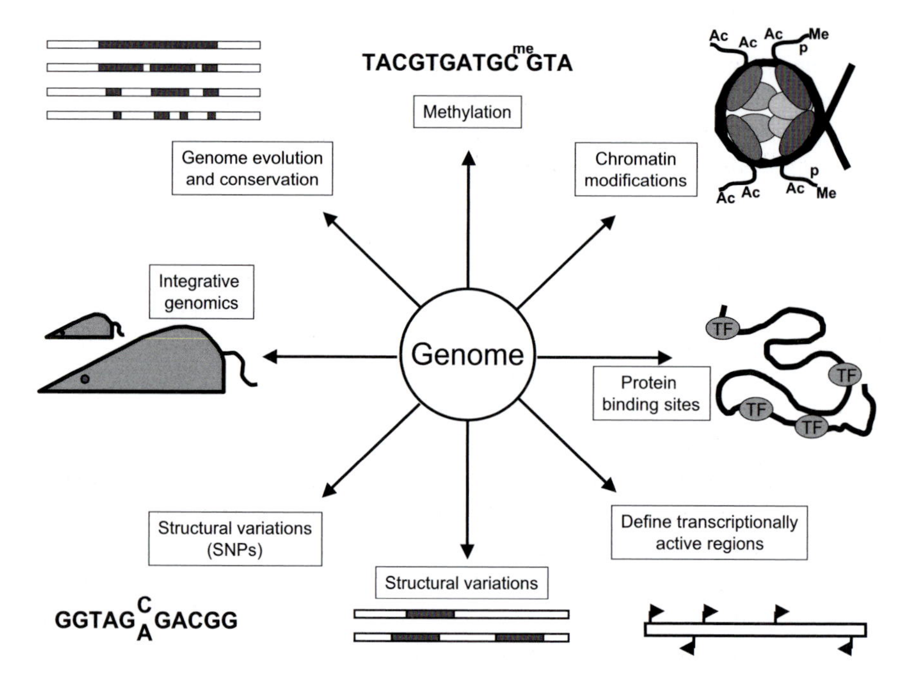

Figure 8.1 DNA microarrays can be used to explore a large spectrum of features of the genome: methylation state and modifications of chromatin (epigenomcs), interactions between proteome and genome, discover and examine transcriptionally active regions, determine structural variations like amplifications/deletions and polymorphisms, link phenotype and genetics through integrative genomics, analyze evolution and conservation of genomes.

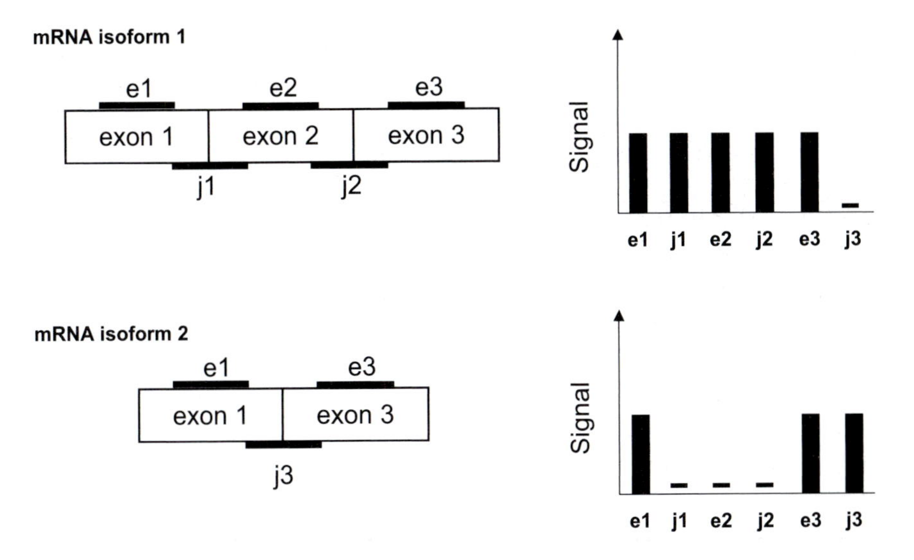

Figure 8.2 Example illustrating how exon and junction probes can be used to monitor mRNA splice variants. If isoform 1 is present all probes except junction probe j3 produce a signal on the array. For isoform 2 probes j1, e2 and j2 lack a signal and junction probe j3 is instead detected. When both isoforms are present a more complicated signal pattern will be generated.

using bridging junction probes. More importantly, also known or predicted splice variants involving non consecutive exons can be monitored using appropriately designed junction probes. Absence (or low abundance) of a particular exon will diminish the exon-specific signal but also modulate the signal for the corresponding junction-probes. Likewise, the presence of a "rare" exon would be indicated by a corresponding increase of a generally low signal. Since several transcripts variants can occur contemporaneously, the overall expression pattern will in general be a superposition of individual variants and can be difficult to interpret. Dominant variants, for example present in a particular tissue or in particular disease sample, can however be detected due to the absence/presence of exon and junction probe intensities relative to the overall gene model. Validation of variants by PCR using appropriately designed primers will in general be necessary. A particularly broad and instructive application of this approach has been performed by Johnson and co-workers (Johnson *et al.*, 2003) Using junction probes for more than 10,000 human genes they surveyed 52 tissues and cell lines for the presence of new and known splice variants. A large number of previously unknown variants was detected by the array and PCR validation for a subset of predicted variants gave a high level of confirmation (Johnson *et al.*, 2003). Similar results were obtained from a detailed analysis using high-density oligonucleotide arrays spanning human chromosomes 21 and 22 (Kapranov *et al.*, 2002; Kampa *et al.*, 2004). Only a small portion of genes (about 21%) was detected with a single isoform. The same study also showed that roughly half of the detected transcripts was outside annotated regions of the genome. Analysis of several thousand mouse alternative transcripts using six probes per alternative splice event again revealed a similar highly complex picture of transcription across tissues (Pan *et al.*, 2004). The same microarray design was subsequently used to address the question if nonsense-mediated mRNA decay regulates gene expression by targeting splice variants containing premature termination codons (Pan *et al.*, 2006).

Other studies have focused on subset of genes or even examined splice variants from a single gene. Fehlbaum and colleagues (Fehlbaum *et al.*, 2005) used five exon and junction probes defining a particular splice event to monitor 43 isoforms from 18 genes. A dedicated microarray containing a preselected set of 364 prostate cancer related genes (1,532 mRNA splice variants) was instead used to generate variant signatures, classify normal and tumor samples and to identify two sub-clusters within the tumor sample group (Zhang *et al.*, 2006). A subset of 64 cancer-related genes was examined in human Breast cancer cell lines using again a splice event sensitive microarray (Srinivasan *et al.*, 2005; Li *et al.*, 2006). Even single genes, for example PTCH (Nagao *et al.*, 2005), are being examined. Tissue- and disease-specific splice variants were examined using both exon and junction probes as well as probes targeting intronic regions (Nagao *et al.*, 2005).

A major step forward to monitor transcript variants such as alternative splice variants but also alternative transcriptional start sites will be the availability of high-density exon-focused microarrays containing as much as 1.4 million probe sets that are becoming available for several organisms (http://www.affymetrix.com/products/arrays/specific/exon.affx). Complementary to this genome-wide platform, smaller customizable gene-family focused arrays will allow to ask very specific questions in a wide variety of applications (http://www.exonhit.com/html/SpliceArray/SpliceArraysGeneFamilies.htm).

From an experimental point of view it should be noted that all these approaches require full-length mRNA as sample material and therefore special amplification protocols (Castle *et al.*, 2003). Many conventional microarray platforms and amplification protocols use amplified transcripts restricted to or biased towards the end of the 3′ untranslated region, often making use of polyT priming for amplification. In particular transcripts lacking a polyA are therefore not amplified using the canonical approach but would be monitored using adequate full-length amplification protocols.

Genomic tiling: transcription outside annotated regions of the genome

From a superficial point of view most of the transcriptional activity could be expected to reside within gene-coding regions. However there is now clear evidence that much larger portions of genomes are actively transcribed (Johnson *et al.*, 2005). This is not only true for yeast (Ross-Macdonald *et al.*, 1999) but even more for complex genomes like the human half of which represents transcriptionally functional units (Semon and Duret, 2004). To examine this phenomenon in more detail so-called genomic tiling techniques have been developed. Rather than focusing only on known or predicted exon containing regions coding for genes, the whole genome (or chromosomes) are systematically scanned, with some restriction in regions where repeat elements are present. Probes are positioned at regular and often overlapping intervals thus allowing detection of transcript information at high resolution (Fig. 8.3).

Initially applied to smaller genomes such as those of *E. coli* and *Saccharomyces cerevisiae* (Ross-Macdonald *et al.*, 1999; Selinger *et al.*, 2000) a large number of genomes or chromosomes from diverse organisms like *Drosophila melanogaster, Arabidopsis* or the human genome have now been surveyed (Selinger *et al.*, 2000; Shoemaker *et al.*, 2001; Kapranov *et al.*, 2002; Rinn *et al.*, 2003; Schadt *et al.*, 2003; Bertone *et al.*, 2004; Schadt *et al.*, 2004; Stolc *et al.*, 2004). In all cases specifically designed tiling microarrays were used containing up to 50 million oligonucleotide probes distributed in regular intervals across the region examined. One important message that became apparent from all these studies was the compelling experimental evidence for a much more complex transcriptome as previously expected. For example almost ten times as much transcription was observed on human chromosomes 21 and 22 than accounted for by transcripts in public databases (Kapranov *et al.*, 2002).

More specific questions, for example the temporal profile of replication of human chromosomes, have also been approached with tiling arrays (Jeon *et al.*, 2005). Methods are continuously improving and transcriptional maps of ten human chromosomes have now been obtained at a five nucleotide resolution

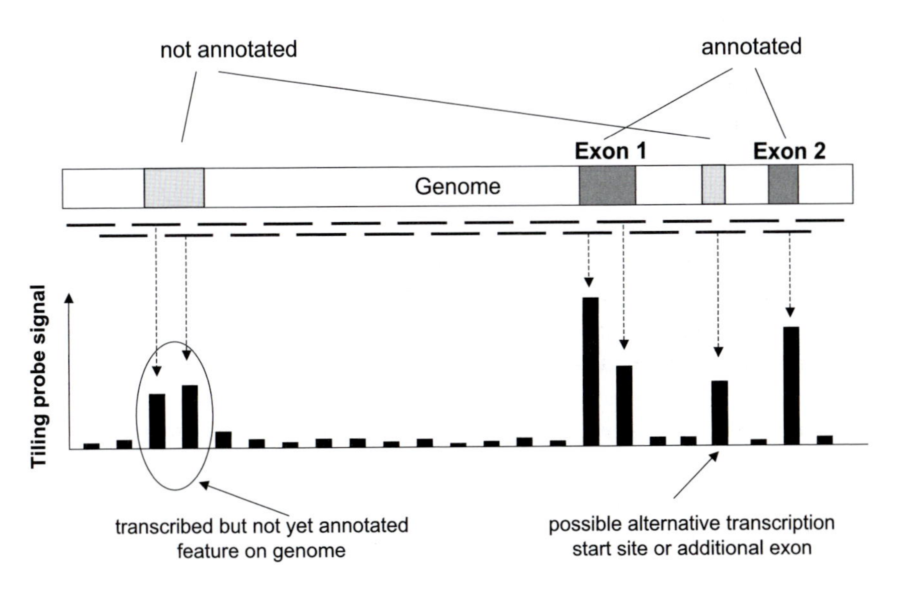

Figure 8.3 In a genomic tiling array the probes are placed at regular intervals along the genome. Probes in transcriptionally active regions show higher signals.

(Cheng *et al.*, 2005). Importantly, a major portion of observed transcripts were non-polyadenylated. At the same time also improvements in sample preparation techniques contribute to an increasingly refined view of the transcriptome (Kapranov *et al.*, 2005).

Profiling of non-coding RNAs

Non-coding RNAs (ncRNAs) or small RNAs represent a large and diverse class of transcripts and are found both in bacteria and higher organisms. Their role and importance in many diverse cellular processes is now widely recognized (Eddy, 2001; Storz, 2002; Ambros, 2004; Bartel, 2004; Huttenhofer *et al.*, 2005; Mattick and Makunin, 2006) making them a target of intense experimental and theoretical research. A number of strategies has been developed to survey the expression and function of known ncRNAs and to discover new candidate ncRNAs genes. Estimates based on bioinformatics analysis of genomes range in the thousands for eukaryotic and hundreds for bacterial genomes (Hershberg *et al.*, 2003; Bartel, 2004; Zhang *et al.*, 2004; Washietl *et al.*, 2005a; Washietl *et al.*, 2005b). The genome-wide tiling experiments mentioned in the previous paragraph have confirmed this view. Data obtained from these and previous studies has been incorporated into prediction algorithms (Bentwich, 2005; Grun *et al.*, 2005; Krek *et al.*, 2005; Washietl *et al.*, 2005b; Rajewsky, 2006; Tjaden *et al.*, 2006; Wang, 2006). New families of ncRNAs are continuously discovered, one of the latest examples being the discovery of a large number of piRNAs (Aravin *et al.*, 2006; Girard *et al.*, 2006; Lau *et al.*, 2006). Given the growing number of distinct ncRNA families a dedicated database, RFAM, has been established (Griffiths-Jones *et al.*, 2005).

DNA microarrays offer a straightforward method to survey the expression of a large number of confirmed or predicted ncRNAs (Huttenhofer and Vogel, 2006; Mattick and Makunin, 2006; Ravasi *et al.*, 2006). *E. coli* was one of the first organisms in which a whole-genome tiling approach was used to examine transcriptionally active intergenic regions (IGRs). A high-density tiling array containing 300,000 probes was designed having a 30 bp resolution in known mRNA, tRNA and rRNA regions and 6 bps in IGRs (Selinger 2000). Subsequent studies (Wasserman *et al*, 2001, Tjaden 2001) then used the same array to more closely examine the IGRs. Variation of growth conditions allowed them to detect signals for a large portion of candidate ncRNAs and also detect and verify a set of new ncRNAs.

Another model organism, the yeast *Saccharomyces cerevisiae*, has also been extensively studied. A microarray containing 212 coding and non-coding RNAs was used to examine several hundred yeast genes with unknown function (Peng *et al.*, 2003). Monitoring RNA profiles allowed many protein-coding genes to be associated with RNA processing or non-coding RNA biogenesis. Using canonical microarrays monitoring mRNA transcripts of yeast mutants a connection between small nucleolar RNAs and exosome function could be established

(Houalla *et al.*, 2006). A microarray containing probes for 161 ncRNAs and 205 mRNAs was instead used to monitor their expression during development and explore stage-specific functions (He *et al.*, 2006).

Expression analysis of 1,602 mouse ncRNAs across 20 tissues revealed a tissue-specific and dynamic expression pattern depending also on external stimuli (Ravasi *et al.*, 2006). Using a microarray containing probes from 3,478 intergenic and intronic sequences conserved between the mouse, rat and human genome Babak and colleagues identified several new ncRNAs by profiling 16 different mouse tissues (Babak *et al.*, 2005), none of which was however conserved across all three genomes.

A particular class of ncRNAs, microRNAs, are now being studied extensively due to their important role in many critical biological processes (Ambros, 2004; Bentwich *et al.*, 2005). Tissue-specific expression of microRNAs has now been clearly documented by a large number of expression profiling studies using amongst other techniques also dedicated microarrays (Barad *et al.*, 2004; Liu *et al.*, 2004; Miska *et al.*, 2004; Monticelli *et al.*, 2005). To further explore the precise biological function of individual microRNAs standard microarrays are instead used. Changes in mRNA expression profiles following transfection with microRNA have been observed and correlate with the tissue-specificity of the transfected microRNA (Lim *et al.*, 2005; Lall *et al.*, 2006; Sood *et al.*, 2006). Analysis of the 3′ untranslated regions of target genes revealed a common signature (Lim *et al.*, 2005) and theoretical efforts to predict microRNA target genes are now an area of extensive research (Bentwich, 2005; Wang, 2006). Another study showed that, like normal mRNA, also microRNAs are subject to changes in expression when cells are treated with histone deacetylase inhibitors, known to affect global transcription levels (Scott *et al.*, 2006). The observation that microRNAs show a tissue-specific expression profile and can effect gene expression has raised a great interest in the field of cancer. Profiling of microRNA is being used to classify cancer (Volinia *et al.*, 2006; Yanaihara *et al.*, 2006) and, like normal mRNA expression profiles, has been suggested as a tool for prognosis prediction (Calin *et al.*, 2005; Lu *et al.*, 2005; Mattie *et al.*, 2006).

Profiling and discovery of microRNAs is an active field including many other organisms like plants (Yamada *et al.*, 2003; Meyers *et al.*, 2006) and viruses (Grundhoff *et al.*, 2006). High-density tiling-style microarrays and the very flexible ink-jet technology have allowed considerable enlargement of our view of the transcriptome encompassing also ncRNAs. Together with new technologies like LNA-based arrays (Castoldi *et al.*, 2006) these array-based approaches will continue to play a central role in this rapidly expanding field.

DNA methylation

Although covalent modifications of genomic DNA has been known for a long time, only the recent development of high-throughput methods have allowed to obtain a detailed genome-wide picture of this phenomenon and examine it under many different biological conditions. A new field, epigenomics (Callinan and Feinberg, 2006; Wilson *et al.*, 2006), encompassing epigenetic marks and other modifications affecting chromatin structure and supra-structure is rapidly expanding. A number of techniques have been developed to perform genome-wide methylation studies (van Steensel, 2005; Schumacher *et al.*, 2006) and map the methylome (Wilson *et al.*, 2006). Comparative genome hybridization (CGH) based approaches (Fig. 8.4) were some of the first array-based methods to detect global methylation patterns of larger genomic regions (Yan *et al.*, 2001; Balog *et al.*, 2002; Gitan *et al.*, 2002; Hatada *et al.*, 2002; Chen *et al.*, 2003; van Steensel and Henikoff, 2003). In array-CGH fragments of genomes to be compared are hybridized onto arrays containing probes derived from BAC clones, cDNAs, PCR products or oligonucleotides (Pinkel and Albertson, 2005) (Fig. 8.4). Depending on the type and number of the probes present on the array different levels of resolution can be obtained. Use of 60- to 100-bp-long oligonucleotides approximately equally distributed along the genome allow for 30–50 kB resolution and further improvements can be expected as array size increase (Barrett *et al.*, 2004; Selzer *et al.*, 2005). Discrimination for methylation is achieved by using restriction enzymes with a methylation-sensitive recognition site and subsequent hybridization to arrays (Fig. 8.5). Comparison with a

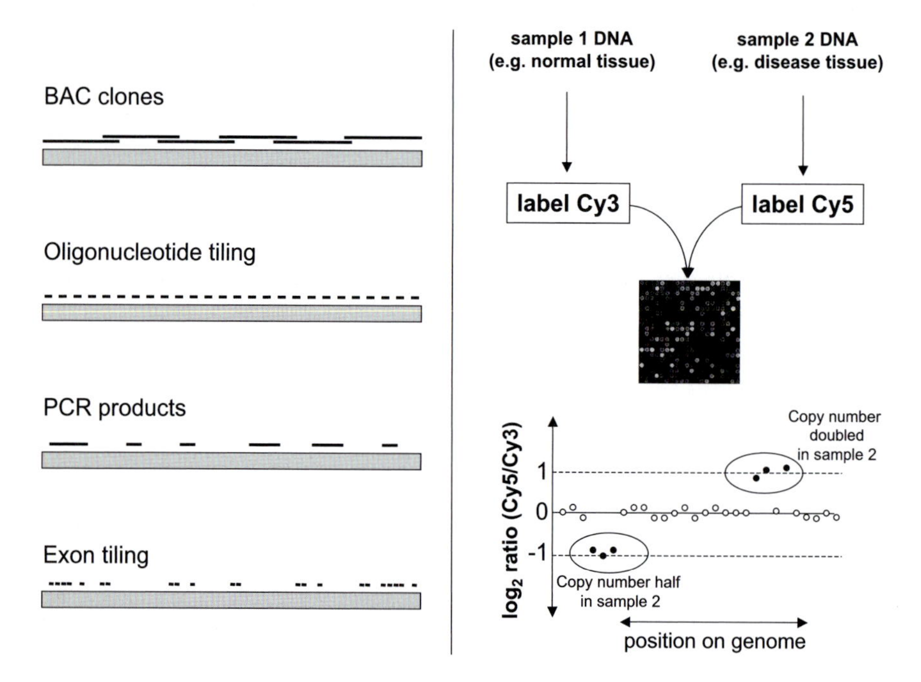

Figure 8.4 For array comparative genome hybridization (array-CGH) probes can be large size BAC clones, 60- to 100-bp-long oligonucleotides for tiling-style layouts, PCR products from selectively amplified regions or probes centered on exons. Comparison of two samples on a two-channel microarray generates a ratio value that represents the relative difference in copy-number for each of the probes.

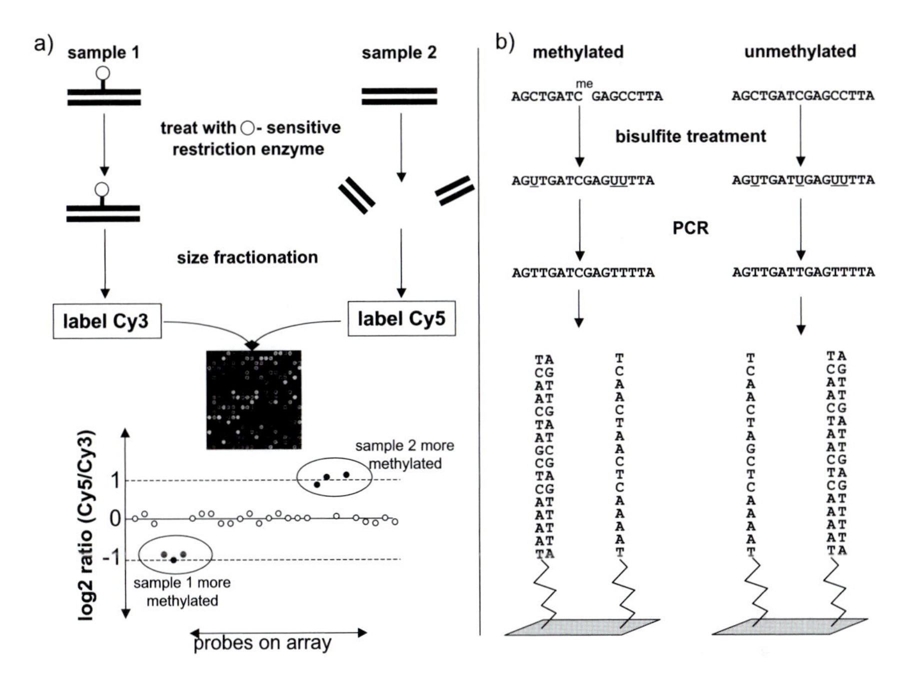

Figure 8.5 (a) To determine differences between the methylation state of two DNA samples the fragments containing unmethylated sites are selectively depleted using methylation-sensitive restriction enzymes. DNA fragments containing unmethylated DNA are cleaved and depleted from the sample. Methylated fragments instead are not affected by the treatment resulting in an enrichment of the latter. (b) Direct readout of the methylation state by bisulfite treatment. On treatment all unmethylated Cs are converted into U which is then converted into T during PCR amplification. By placing probes on the array that allow to discriminate between a C and T at the position under question the methylation state can be determined.

reference sample then allows to detect enrichment or depletion of methylation across large regions (van Steensel and Henikoff, 2003). Focus on more restricted regions of methylation has been performed using a bisulfite-based strategy and hybridization conditions that allow to discriminate between C or T at a methylatable position annealing only to the correct oligonucleotide on the array (van Steensel and Henikoff, 2003). In contrast to the CGH-type method, which compares the global methylation states of samples, this method allows a direct readout of the methylation state in a particular sample (Fig. 8.5). Techniques in sample preparation are also continuously improving, for example by enrichment of unmethylated fractions that increases the sensitivity for detection of new methylation sites (Schumacher *et al* 2006) or the use of an immunoprecipitation procedure recognizing methylated CpG DNA (Weber *et al.*, 2005; Gebhard *et al.*, 2006). Promoter regions are frequently enriched in CpG islands and therefore represent major targets for methylation. A number of studies has therefore focused on promoter methylation state of diseased and normal tissue (Leu *et al.*, 2004; Ching *et al.*, 2005; Costello, 2005; Weber *et al.*, 2005; Fukasawa *et al.*, 2006; Gebhard *et al.*, 2006; Hatada *et al.*, 2006; Keshet *et al.*, 2006; Wilson *et al.*, 2006). Microarray-derived methylation patterns have also been used to classify tumors (Adorjan *et al.*, 2002) or predict prognosis (Wei *et al.*, 2002).

"ChIP-on-chip" for mapping protein–DNA interactions and epigenetic marks

The availability of a steadily increasing number of fully sequenced genomes combined with large volumes of transcript data has allowed obtaining a rather detailed picture which genes are actively transcribed under specific conditions. Mapping this information onto the genome allowed identification of regulatory regions such as promoters. A deeper understanding how the transcriptional activity of a particular gene is regulated requires however also knowledge which regulatory factors are present under specific conditions.

Using DNA microarrays this question is being approached with the help of a technique called chromatin immunoprecipitation-on-a-chip (ChIP-on-chip) (Fig. 8.6). One of the first application of this method has been the identification of genome-wide binding sites of various yeast transcription factors (Ren *et al.*, 2000; Iyer *et al.*, 2001; Lieb *et al.*, 2001).

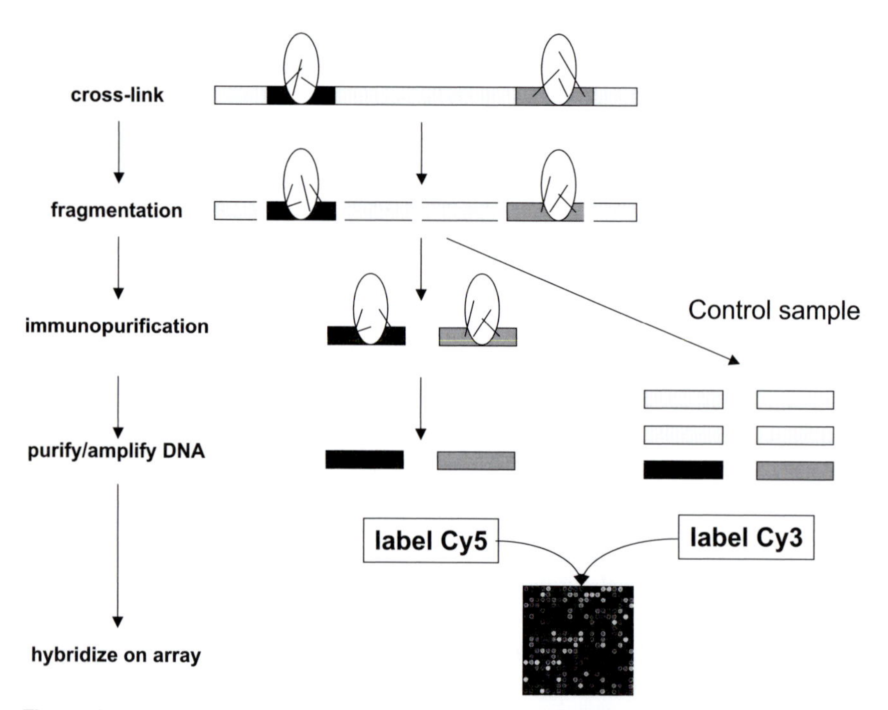

Figure 8.6 Principle of ChIP-on-chip. Proteins are cross-linked to genomic DNA after which samples are fragmented into small pieces. Immunopurification using an antibody specific for a protein or a covalent modification of the protein isolates only DNA fragments bound by the protein. Comparison with an untreated control sample indicates regions of the genome with bound factors (or specifically modified proteins like histones).

The basic principle behind this method relies on cross-linking of proteins to genomic DNA. Following cross-linking of DNA-bound proteins the DNA is mechanically broken up into small fragments (0.2–2.0 kB) using sonication. Specific protein–DNA complexes can now be purified using immunoprecipitation by one or more antibodies recognizing the protein(s) of interest. After reversing the cross-link the purified DNA can now be further processed (purification, PCR amplification, labeling) for examination on a DNA microarray (Rodriguez and Huang, 2005; van Steensel, 2005; Orian, 2006).

A number of strategies have been developed to design probes for DNA microarrays suitable for ChIP-on-chip analyses: high-density oligonucleotides covering genomic regions of interest (Kapranov *et al.*, 2002), PCR fragments of target genes (Li *et al.*, 2003) or CpG-island (CGI) clone libraries (Heisler *et al.*, 2005). Depending on the question to be addressed, these methods allow to perform global surveys or focus more specifically on particular region of interest.

For example, high density oligonucleotide arrays have been designed to examine transcription factor (TF) binding sites (Sp1, c-Myc, and p53) across all non-repetitive sequences of human chromosomes 21 and 22 (Cawley *et al.*, 2004). This approach has been further extended resulting in the availability of commercial oligonucleotide arrays (AFFYMETRIX) that allow scanning of the whole human and mouse genome. Using a PCR based approach to generate customizable DNA arrays Li and colleagues have examined promoters of about 5,000 human genes to identify the genomic binding sites of c-Myc in Burkitt's lymphoma cells (Li *et al.*, 2003). Using the same approach arrays can be generated that contain a contiguous tile-path that "walks" across one or more chromosomes, for example to map protein-DNA interactions in *Drosophila* across large genomic regions (Li *et al.*, 2003). Use of this PCR-approach is particularly attractive for organisms where no commercial arrays are (yet) available or where a focus on a specific genomic region is required.

A third way to construct microarrays for CHiP analysis is to use CGI focused libraries taking advantage of the strong association between CGIs and promoters or gene regulatory regions. An array containing 7,776 CGI clones from the CGI genomic library of the United Kingdom Genome Mapping Project Centre (http://www.sanger.ac.uk/HGP/cgi.shtml) has been used to compare hyper methylation in breast tumor and normal samples (Yan *et al.*, 2004). A human CGI array containing 12,192 CGI clones prepared from the same genomic library is now available from the Ontario University Health Network. (www.microarrays.ca). More recently the number of clones has been further extended and now covers around 9600 genomic loci across all chromosomes (Heisler *et al.*, 2005).

In addition to the aforementioned Affymetrix array also Agilent now has a "Human ENCODE ChIP-on-chip Microarray Set" specifically designed for ChIP-on-Chip analysis. The two-slide set contains over 80,000 probes that cover human chromosomes 1 to 22 within specific ENCODE regions (http://www.chem.agilent.com). Likewise, Nimblegen also offers various oligonucleotide-based arrays with up to 400,000 probes (http://www.nimblegen.com/products/cgh/index.html).

While initially focusing on general protein-DNA interactions ChIP-on-chip applications are increasingly applied to another, related field, Epigenomics focusing on modifications altering chromatin structure. In particular, covalent modification of residues in histone tails (acetylation, methylation, phosphorylation) has been shown to be intimately connected to chromatin structure and transcription (Huebert and Bernstein, 2005; Rodriguez and Huang, 2005; Orian, 2006; Saha *et al.*, 2006). ChIP-on-chip using antibodies specific for a particular modification has been instrumental to the progress in this field during the recent years. In addition to a genome-wide view of histone modification ChIP-on-chip can be combined with a parallel standard expression analysis to derive correlated changes in gene expression. Initial analysis in yeast has shown that histone H3 and H4 acetylation and methylation within promoter regions and H3–K4 dimethylation in coding regions significantly correlates with transcriptional activity (Bernstein and Schreiber, 2002; Bernstein *et al.*, 2004; Huebert and Bernstein, 2005). Similar results have been observed in Drosophila showing again a binary pattern of modifications in euchromatic genes (Schubeler *et al.*, 2004). As for general TF binding site analysis, high-density oligonucleotide arrays are now being routinely used to construct genome-wide maps of histone modifications, for example H3–K4 di- and tri-methylation and H3–K9/14 acetylation (Huebert and Bernstein, 2005).

Covalent modification of histones and global chromatin accessibility are directly interconnected and, using appropriate purification protocols, chromatin can be separated by its condensation state. Using a chromatin solubility assay, mono- and oligonucleosomes can be separated into three fractions—S1, S2, and P—with S2 representing the most condensed chromatin fraction (Rodriguez and Huang, 2005). Comparing the S2 fraction with a total genomic DNA reference on a DNA microarray covering the complete genome the condensation state of thousands of individual loci can be determined and correlated with measurements regarding their transcriptional activity (Weil *et al.*, 2004). This chromatin array method is therefore unique because it allows to examine epigenetic regulation at a higher level of chromatin state as compared to standard methods.

While ChIP-on-chip based methods now play a central role in exploring the epigenetic landscape of the genome (Carroll *et al.*, 2005; Fraga and Esteller, 2005; Huebert and Bernstein, 2005; Pokholok *et al.*, 2005; Orian, 2006) the method can be generally applied to any DNA-binding protein. Examples are mapping of estrogen-receptor binding sites (Carroll *et al.*, 2005), mapping of multiple transcription factor binding sites across chromosomes 21 and 22 (Cawley *et al.*, 2004), genome-wide detection of polycomb targets (Schwartz *et al.*, 2006) or transcription factor binding sites (Orian *et al.*, 2003; Cawley *et al.*, 2004) in Drosophila.

A technique complementary to ChIP is DNA adenine transferase identification (DamID), which also allows mapping of DNA–protein interactions through DNA microarrays (van Steensel and Henikoff, 2000; van Steensel, 2005; Orian, 2006). Fusion of Dam to a DNA-binding protein introduces unique methylation marks on adenine bases close to the binding site of the protein that can then be recognized and cleaved by specific restriction enzymes. Similar to studies examining the methylation status of the genome, a second sample containing only the native DNA-binding protein without Dam (i.e., no methylation) is used as reference. The two samples are then compared on a two-channel microarray and binding sites are indicated by differences in spot intensity (van Steensel *et al.*, 2001; Orian *et al.*, 2003; Song *et al.*, 2004)

Detection of genome variations

Structural variations in the human (and other) genomes have been known for a long time. Development of high-throughput technologies utilizing DNA microarrays have now become available allowing a detailed genome-wide analysis: array-CGH and SNP arrays (Syvanen, 2005; van den Ijssel *et al.*, 2005; Feuk *et al.*, 2006; Ylstra *et al.*, 2006). In addition special sample preparation methods like ROMA (Kennedy *et al.*, 2003; Lucito *et al.*, 2003) improve sample quality and give better signal-to-noise ratios. For array-CGH copy-number changes (amplifications or deletions) are directly measured due to the presence of a second control sample in the hybridization.

SNP arrays instead take advantage of the fact that on average one single nucleotide polymorphism (SNP) is present every 300 nucleotides. Arrays comprising 10K up to 100K or even 500K probe-sets are now available allowing sampling at various levels of resolution (Dong *et al.*, 2001; Warrington *et al.*, 2002; Matsuzaki *et al.*, 2004a; Matsuzaki *et al.*, 2004b; Hardenbol *et al.*, 2005; Herr *et al.*, 2005; Hinds *et al.*, 2005). For yeast an array allowing comparison of different strains at single nucleotide resolution has recently been described (Gresham *et al.*, 2006). For SNP arrays, copy numbers need to be derived by comparison to averages obtained from separate experiments on control samples. On the other hand, one advantage of SNP arrays with respect to array-CGH is their capability to detect also loss of heterozygosity (Matsuzaki *et al.*, 2004a; Hardenbol *et al.*, 2005). Additional platforms are actively being developed (Nuwaysir *et al.*, 2002; Gunderson *et al.*, 2005) and will allow an even greater choice of methods in the future.

During the past few years both these technologies have allowed to dramatically increase the extend and detail of structural variations of the human genome (Huang *et al.*, 2004; Iafrate *et al.*, 2004; Ishkanian *et al.*, 2004; Sebat *et al.*, 2004; Buckley *et al.*, 2005; Miller *et al.*, 2005; Sharp *et al.*, 2005; Slater *et al.*, 2005;

Tuzun *et al.*, 2005; Eichler, 2006; Perry *et al.*, 2006), explore variations in cancer genomes (Hodgson *et al.*, 2001; Massion *et al.*, 2002; Albertson and Pinkel, 2003; Lucito *et al.*, 2003; Martinez-Climent *et al.*, 2003; Bignell *et al.*, 2004; Brennan *et al.*, 2004; Zhao *et al.*, 2004; Benetkiewicz *et al.*, 2005; Cox *et al.*, 2005; Gunderson *et al.*, 2005; Mestre *et al.*, 2005; Pinkel and Albertson, 2005; Engle *et al.*, 2006) or, more general, correlate structural genomic variations to disease (Mantripragada *et al.*, 2004; Sharp *et al.*, 2005; Vissers *et al.*, 2005; Herbert *et al.*, 2006). Influence of copy-number variations on drug response is another hypothesis that is now being examined (Ouahchi *et al.*, 2006).

Integrative genomics

The final goal of many "omics-" centered research aims at the elucidation of complex gene interaction networks that give rise to a particular global phenotype, for example a disease phenotype. A central problem is the distinction between causative and reactive components in the underlying network which is difficult if not impossible to achieve using gene expression data alone. By adding genetic information a combined genome-wide genetic analysis of gene expression is however possible as initially demonstrated in budding yeast (Jansen and Nap, 2001; Brem *et al.*, 2002). Using some of the microarray platforms described in the previous paragraphs this approach has recently been applied also to complex traits in higher organisms (Schadt *et al.*, 2003; Chesler *et al.*, 2005; Schadt, 2005; Schadt *et al.*, 2005; Hubner *et al.*, 2006; Lum *et al.*, 2006). Genome-wide SNP genotyping and expression profiling of a population of segregating mice or rats allowed to discover new genes related to obesity (Schadt *et al.*, 2005), hypertension (Hubner *et al.*, 2006), neural synapse function (Chesler *et al.*, 2005) or diabetes (Lum *et al.*, 2006). Given the power of this integrative genomic approach many more applications in dissecting disease or drug response (Schadt, 2005) can be expected and initiatives like the "Collaborative Cross" will further enhance this trend by providing a large common set of genetically defined mice (Churchill *et al.*, 2004).

Microarrays focusing on viruses and bacteria

Genomes of higher organisms like man, mouse or plants are highly complex and therefore are primary targets for the application of DNA microarrays. However, also the analysis of smaller genomes like those of bacteria and viruses can benefit from this technique. One area is the discovery of new or examination of existing virus genera and strains. Chou and co-workers designed an array containing probes for 5,700 different viruses (Chou *et al.*, 2006). The array contains both probes conserved within a viral genus and also strain-specific probes. This allows the identification of existing viruses and also new emerging viral strains can be detected (Chou *et al.*, 2006). A similar approach has been followed by Wang and colleagues (Wang *et al.*, 2002) allowing them to readily detect a new member of the coronavirus family during the 2003 SARS outbreak (Wang *et al.*, 2002). For influenza A virus a oligonucleotide microarray has been designed to perform rapid subtype identification and sequencing (Lodes *et al.*, 2006). Another project, called "LAB ON A CHIP," is focused on the development of a DNA microarray able to rapidly detect more than 600 dangerous animal viruses (http://www.defra.gov.uk/news/2006/060523b.htm). Microarrays containing the whole genome of the plague causing *Yersinia pestis* are used to examine the genomic content of live plague vaccine (Zhou *et al.*, 2004). A related study compared genomes from various strains of *Yestinia pestis* and the related *Yersinia pseudotuberculosis* (Hinchliffe *et al.*, 2003). Comparative phylogenomics of bacteria using microarrays has been applied to a wide spectrum of cases and has allowed to obtain new insights into virulence and evolution of bacterial pathogens (Albert *et al.*, 2005; Dorrell *et al.*, 2005; Hamelin *et al.*, 2006; Raskin *et al.*, 2006). Other examples are detection of pathogen populations in clinical samples (Lin *et al.*, 2006) or live food (Kostrzynska and Bachand, 2006).

Conclusions

As outlined in the preceding paragraphs DNA microarray technology can be applied in the context of a wide range of biological and biomedical problems across practically all organisms. As available technologies and analysis methods develop further and more genomes become annotated the use of DNA microarrays will steadily increase and become a routine tool in modern biology, or more general, in biomedical research and diagnosis.

References

Adorjan, P., Distler, J., Lipscher, E., Model, F., Muller, J., Pelet, C., Braun, A., Florl, A. R., Gutig, D., Grabs, G., *et al.* (2002). Tumour class prediction and discovery by microarray-based DNA methylation analysis. Nucleic Acids Res. *30*, e21.

Albert, T.J., Dailidiene, D., Dailide, G., Norton, J.E., Kalia, A., Richmond, T. A., Molla, M., Singh, J., Green, R.D., and Berg, D.E. (2005). Mutation discovery in bacterial genomes: metronidazole resistance in Helicobacter pylori. Nat. Methods 2, 951–953.

Albertson, D.G., and Pinkel, D. (2003). Genomic microarrays in human genetic disease and cancer. Hum. Mol. Genet. *12 Spec. No. 2*, R145–152.

Ambros, V. (2004). The functions of animal microRNAs. Nature *431*, 350–355.

Aravin, A., Gaidatzis, D., Pfeffer, S., Lagos-Quintana, M., Landgraf, P., Iovino, N., Morris, P., Brownstein, M. J., Kuramochi-Miyagawa, S., Nakano, T., *et al.* (2006). A novel class of small RNAs bind to MILI protein in mouse testes. Nature.

Babak, T., Blencowe, B. J., and Hughes, T. R. (2005). A systematic search for new mammalian noncoding RNAs indicates little conserved intergenic transcription. BMC Genomics 6, 104.

Balog, R.P., de Souza, Y.E., Tang, H.M., DeMasellis, G.M., Gao, B., Avila, A., Gaban, D. J., Mittelman, D., Minna, J.D., Luebke, K.J., and Garner, H.R. (2002). Parallel assessment of CpG methylation by two-color hybridization with oligonucleotide arrays. Anal. Biochem. *309*, 301–310.

Barad, O., Meiri, E., Avniel, A., Aharonov, R., Barzilai, A., Bentwich, I., Einav, U., Gilad, S., Hurban, P., Karov, Y., *et al.* (2004). MicroRNA expression detected by oligonucleotide microarrays: system establishment and expression profiling in human tissues. Genome Res. *14*, 2486–2494.

Barrett, M.T., Scheffer, A., Ben-Dor, A., Sampas, N., Lipson, D., Kincaid, R., Tsang, P., Curry, B., Baird, K., Meltzer, P.S., *et al.* (2004). Comparative genomic hybridization using oligonucleotide microarrays and total genomic DNA. Proc. Natl. Acad. Sci. USA *101*, 17765–17770.

Bartel, D.P. (2004). MicroRNAs: genomics, biogenesis, mechanism, and function. Cell *116*, 281–297.

Benetkiewicz, M., Wang, Y., Schaner, M., Wang, P., Mantripragada, K.K., Buckley, P.G., Kristensen, G., Borresen-Dale, A.L., and Dumanski, J.P. (2005). High-resolution gene copy number and expression profiling of human chromosome 22 in ovarian carcinomas. Genes Chromosomes Cancer *42*, 228–237.

Bentwich, I. (2005). Prediction and validation of microRNAs and their targets. FEBS Lett. *579*, 5904–5910.

Bentwich, I., Avniel, A., Karov, Y., Aharonov, R., Gilad, S., Barad, O., Barzilai, A., Einat, P., Einav, U., Meiri, E., *et al.* (2005). Identification of hundreds of conserved and nonconserved human microRNAs. Nat. Genet. *37*, 766–770.

Bernstein, B.E., Humphrey, E.L., Liu, C.L., and Schreiber, S.L. (2004). The use of chromatin immunoprecipitation assays in genome-wide analyses of histone modifications. Methods Enzymol. *376*, 349–360.

Bernstein, B.E., and Schreiber, S.L. (2002). Global approaches to chromatin. Chem. Biol. *9*, 1167–1173.

Bertone, P., Stolc, V., Royce, T.E., Rozowsky, J.S., Urban, A.E., Zhu, X., Rinn, J.L., Tongprasit, W., Samanta, M., Weissman, S., *et al.* (2004). Global identification of human transcribed sequences with genome tiling arrays. Science *306*, 2242–2246.

Bignell, G. R., Huang, J., Greshock, J., Watt, S., Butler, A., West, S., Grigorova, M., Jones, K. W., Wei, W., Stratton, M.R., *et al.* (2004). High-resolution analysis of DNA copy number using oligonucleotide microarrays. Genome Res. *14*, 287–295.

Bild, A.H., Yao, G., Chang, J.T., Wang, Q., Potti, A., Chasse, D., Joshi, M. B., Harpole, D., Lancaster, J. M., Berchuck, A., *et al.* (2006). Oncogenic pathway signatures in human cancers as a guide to targeted therapies. Nature *439*, 353–357.

Brem, R.B., Yvert, G., Clinton, R., and Kruglyak, L. (2002). Genetic dissection of transcriptional regulation in budding yeast. Science *296*, 752–755.

Brennan, C., Zhang, Y., Leo, C., Feng, B., Cauwels, C., Aguirre, A.J., Kim, M., Protopopov, A., and Chin, L. (2004). High-resolution global profiling of genomic alterations with long oligonucleotide microarray. Cancer Res. *64*, 4744–4748.

Buckley, P.G., Mantripragada, K. K., Piotrowski, A., Diaz de Stahl, T., and Dumanski, J.P. (2005). Copy-number polymorphisms: mining the tip of an iceberg. Trends Genet. *21*, 315–317.

Calin, G.A., Ferracin, M., Cimmino, A., Di Leva, G., Shimizu, M., Wojcik, S.E., Iorio, M.V., Visone, R., Sever, N.I., Fabbri, M., *et al.* (2005). A MicroRNA signature associated with prognosis and progression in chronic lymphocytic leukemia. N. Engl. J. Med. *353*, 1793–1801.

Callinan, P.A., and Feinberg, A.P. (2006). The emerging science of epigenomics. Hum. Mol. Genet. *15 Spec. No. 1*, R95–101.

Carroll, J.S., Liu, X.S., Brodsky, A.S., Li, W., Meyer, C.A., Szary, A.J., Eeckhoute, J., Shao, W., Hestermann, E. V., Geistlinger, T.R., *et al.* (2005). Chromosome-wide mapping of estrogen receptor binding reveals long-range regulation requiring the forkhead protein FoxA1. Cell *122*, 33–43.

Castle, J., Garrett-Engele, P., Armour, C.D., Duenwald, S. J., Loerch, P.M., Meyer, M.R., Schadt, E.E., Stoughton, R., Parrish, M.L., Shoemaker, D.D., and Johnson, J.M. (2003). Optimization of oligonucleotide arrays and RNA amplification protocols for analysis of transcript structure and alternative splicing. Genome Biol. *4*, R66.

Castoldi, M., Schmidt, S., Benes, V., Noerholm, M., Kulozik, A. E., Hentze, M. W., and Muckenthaler, M.U. (2006). A sensitive array for microRNA expression profiling (miChip) based on locked nucleic acids (LNA). RNA *12*, 913–920.

Cawley, S., Bekiranov, S., Ng, H.H., Kapranov, P., Sekinger, E. A., Kampa, D., Piccolboni, A., Sementchenko, V., Cheng, J., Williams, A. J., *et al.* (2004). Unbiased mapping of transcription factor binding sites along human chromosomes 21 and 22 points to widespread regulation of noncoding RNAs. Cell *116*, 499–509.

Chen, C.M., Chen, H. L., Hsiau, T.H., Hsiau, A.H., Shi, H., Brock, G.J., Wei, S. H., Caldwell, C. W., Yan, P.S., and Huang, T.H. (2003). Methylation target array for rapid analysis of CpG island hypermethylation in multiple tissue genomes. Am. J. Pathol. *163*, 37–45.

Cheng, J., Kapranov, P., Drenkow, J., Dike, S., Brubaker, S., Patel, S., Long, J., Stern, D., Tammana, H., Helt, G., *et al.* (2005). Transcriptional maps of 10 human chromosomes at 5-nucleotide resolution. Science *308*, 1149–1154.

Chesler, E.J., Lu, L., Shou, S., Qu, Y., Gu, J., Wang, J., Hsu, H.C., Mountz, J.D., Baldwin, N.E., Langston, M.A., *et al.* (2005). Complex trait analysis of gene expression uncovers polygenic and pleiotropic networks that modulate nervous system function. Nat. Genet. *37*, 233–242.

Ching, T.T., Maunakea, A.K., Jun, P., Hong, C., Zardo, G., Pinkel, D., Albertson, D.G., Fridlyand, J., Mao, J. H., Shchors, K., *et al.* (2005). Epigenome analyses using BAC microarrays identify evolutionary conservation of tissue-specific methylation of SHANK3. Nat. Genet. *37*, 645–651.

Chou, C.C., Lee, T.T., Chen, C.H., Hsiao, H.Y., Lin, Y.L., Ho, M.S., Yang, P.C., and Peck, K. (2006). Design of microarray probes for virus identification and detection of emerging viruses at the genus level. BMC Bioinformatics *7*, 232.

Churchill, G.A., Airey, D.C., Allayee, H., Angel, J.M., Attie, A.D., Beatty, J., Beavis, W.D., Belknap, J.K., Bennett, B., Berrettini, W., *et al.* (2004). The Collaborative Cross, a community resource for the genetic analysis of complex traits. Nat. Genet. *36*, 1133–1137.

Costello, J.F. (2005). Comparative epigenomics of leukemia. Nat. Genet. *37*, 211–212.

Cox, C., Bignell, G., Greenman, C., Stabenau, A., Warren, W., Stephens, P., Davies, H., Watt, S., Teague, J., Edkins, S., *et al.* (2005). A survey of homozygous deletions in human cancer genomes. Proc. Natl. Acad. Sci. USA *102*, 4542–4547.

Dong, S., Wang, E., Hsie, L., Cao, Y., Chen, X., and Gingeras, T.R. (2001). Flexible use of high-density oligonucleotide arrays for single-nucleotide polymorphism discovery and validation. Genome Res *11*, 1418–1424.

Dorrell, N., Hinchliffe, S. J., and Wren, B.W. (2005). Comparative phylogenomics of pathogenic bacteria by microarray analysis. Curr. Opin. Microbiol. *8*, 620–626.

Downward, J. (2006). Cancer biology: signatures guide drug choice. Nature *439*, 274–275.

Eddy, S.R. (2001). Non-coding RNA genes and the modern RNA world. Nat. Rev. Genet. *2*, 919–929.

Eichler, E.E. (2006). Widening the spectrum of human genetic variation. Nat. Genet. *38*, 9–11.

Engle, L.J., Simpson, C.L., and Landers, J.E. (2006). Using high-throughput SNP technologies to study cancer. Oncogene *25*, 1594–1601.

Fehlbaum, P., Guihal, C., Bracco, L., and Cochet, O. (2005). A microarray configuration to quantify expression levels and relative abundance of splice variants. Nucleic Acids Res. *33*, e47.

Feuk, L., Carson, A. R., and Scherer, S. W. (2006). Structural variation in the human genome. Nat. Rev. Genet. *7*, 85–97.

Fraga, M.F., and Esteller, M. (2005). Towards the human cancer epigenome: a first draft of histone modifications. Cell Cycle *4*, 1377–1381.

Fukasawa, M., Kimura, M., Morita, S., Matsubara, K., Yamanaka, S., Endo, C., Sakurada, A., Sato, M., Kondo, T., Horii, A., *et al.* (2006). Microarray analysis of promoter methylation in lung cancers. J. Hum. Genet. *51*, 368–374.

Gebhard, C., Schwarzfischer, L., Pham, T. H., Schilling, E., Klug, M., Andreesen, R., and Rehli, M. (2006). Genome-wide profiling of CpG methylation identifies novel targets of aberrant hypermethylation in myeloid leukemia. Cancer Res. *66*, 6118–6128.

Girard, A., Sachidanandam, R., Hannon, G.J., and Carmell, M.A. (2006). A germline-specific class of small RNAs binds mammalian Piwi proteins. Nature.

Gitan, R.S., Shi, H., Chen, C. M., Yan, P.S., and Huang, T.H. (2002). Methylation-specific oligonucleotide microarray: a new potential for high-throughput methylation analysis. Genome Res *12*, 158–164.

Gresham, D., Ruderfer, D. M., Pratt, S. C., Schacherer, J., Dunham, M.J., Botstein, D., and Kruglyak, L. (2006). Genome-wide detection of polymorphisms at nucleotide resolution with a single DNA microarray. Science *311*, 1932–1936.

Griffiths-Jones, S., Moxon, S., Marshall, M., Khanna, A., Eddy, S.R., and Bateman, A. (2005). Rfam: annotating non-coding RNAs in complete genomes. Nucleic Acids Res. *33*, D121–124.

Grun, D., Wang, Y.L., Langenberger, D., Gunsalus, K.C., and Rajewsky, N. (2005). microRNA target predictions across seven Drosophila species and comparison to mammalian targets. PLoS Comput. Biol. *1*, e13.

Grundhoff, A., Sullivan, C.S., and Ganem, D. (2006). A combined computational and microarray-based approach identifies novel microRNAs encoded by human gamma-herpesviruses. RNA *12*, 733–750.

Gunderson, K.L., Steemers, F. J., Lee, G., Mendoza, L.G., and Chee, M.S. (2005). A genome-wide scalable SNP genotyping assay using microarray technology. Nat. Genet. *37*, 549–554.

Hamelin, K., Bruant, G., El-Shaarawi, A., Hill, S., Edge, T. A., Bekal, S., Fairbrother, J.M., Harel, J., Maynard, C., Masson, L., and Brousseau, R. (2006). A virulence and antimicrobial resistance DNA microarray detects a high frequency of virulence genes in *Escherichia coli* isolates from Great Lakes recreational waters. Appl. Environ. Microbiol. *72*, 4200–4206.

Hanash, S. (2004). Integrated global profiling of cancer. Nat. Rev. Cancer *4*, 638–644.

Hardenbol, P., Yu, F., Belmont, J., Mackenzie, J., Bruckner, C., Brundage, T., Boudreau, A., Chow, S., Eberle, J., Erbilgin, A., *et al.* (2005). Highly multiplexed molecular inversion probe genotyping: over 10,000 targeted SNPs genotyped in a single tube assay. Genome Res. *15*, 269–275.

Hatada, I., Fukasawa, M., Kimura, M., Morita, S., Yamada, K., Yoshikawa, T., Yamanaka, S., Endo, C., Sakurada, A., Sato, M., *et al.* (2006). Genome-wide profiling of promoter methylation in human. Oncogene *25*, 3059–3064.

Hatada, I., Kato, A., Morita, S., Obata, Y., Nagaoka, K., Sakurada, A., Sato, M., Horii, A., Tsujimoto, A., and Matsubara, K. (2002). A microarray-based method for detecting methylated loci. J. Hum. Genet. *47*, 448–451.

He, H., Cai, L., Skogerbo, G., Deng, W., Liu, T., Zhu, X., Wang, Y., Jia, D., Zhang, Z., Tao, Y., *et al.* (2006). Profiling Caenorhabditis elegans non-coding RNA expression with a combined microarray. Nucleic Acids Res. *34*, 2976–2983.

Heisler, L. E., Torti, D., Boutros, P. C., Watson, J., Chan, C., Winegarden, N., Takahashi, M., Yau, P., Huang, T. H., Farnham, P. J., *et al.* (2005). CpG Island microarray probe sequences derived from a physical library are representative of CpG Islands annotated on the human genome. Nucleic Acids Res. *33*, 2952–2961.

Herbert, A., Gerry, N. P., McQueen, M. B., Heid, I. M., Pfeufer, A., Illig, T., Wichmann, H. E., Meitinger, T., Hunter, D., Hu, F. B., *et al.* (2006). A common genetic variant is associated with adult and childhood obesity. Science *312*, 279–283.

Herr, A., Grutzmann, R., Matthaei, A., Artelt, J., Schrock, E., Rump, A., and Pilarsky, C. (2005). High-resolution analysis of chromosomal imbalances using the Affymetrix 10K SNP genotyping chip. Genomics *85*, 392–400.

Hershberg, R., Altuvia, S., and Margalit, H. (2003). A survey of small RNA-encoding genes in Escherichia coli. Nucleic Acids Res. *31*, 1813–1820.

Hinchliffe, S.J., Isherwood, K.E., Stabler, R. A., Prentice, M.B., Rakin, A., Nichols, R.A., Oyston, P. C., Hinds, J., Titball, R.W., and Wren, B.W. (2003). Application of DNA microarrays to study the evolutionary genomics of Yersinia pestis and Yersinia pseudotuberculosis. Genome Res. *13*, 2018–2029.

Hinds, D. A., Stuve, L. L., Nilsen, G. B., Halperin, E., Eskin, E., Ballinger, D. G., Frazer, K. A., and Cox, D. R. (2005). Whole-genome patterns of common DNA variation in three human populations. Science *307*, 1072–1079.

Hodgson, G., Hager, J. H., Volik, S., Hariono, S., Wernick, M., Moore, D., Nowak, N., Albertson, D. G., Pinkel, D., Collins, C., *et al.* (2001). Genome scanning with array CGH delineates regional alterations in mouse islet carcinomas. Nat. Genet. *29*, 459–464.

Houalla, R., Devaux, F., Fatica, A., Kufel, J., Barrass, D., Torchet, C., and Tollervey, D. (2006). Microarray detection of novel nuclear RNA substrates for the exosome. Yeast *23*, 439–454.

Huang, J., Wei, W., Zhang, J., Liu, G., Bignell, G.R., Stratton, M.R., Futreal, P.A., Wooster, R., Jones, K.W., and Shapero, M.H. (2004). Whole genome DNA copy number changes identified by high density oligonucleotide arrays. Hum. Genomics *1*, 287–299.

Hubner, N., Yagil, C., and Yagil, Y. (2006). Novel integrative approaches to the identification of candidate genes in hypertension. Hypertension *47*, 1–5.

Huebert, D.J., and Bernstein, B.E. (2005). Genomic views of chromatin. Curr. Opin. Genet. Dev. *15*, 476–481.

Huttenhofer, A., Schattner, P., and Polacek, N. (2005). Non-coding RNAs: hope or hype? Trends Genet. *21*, 289–297.

Huttenhofer, A., and Vogel, J. (2006). Experimental approaches to identify non-coding RNAs. Nucleic Acids Res. *34*, 635–646.

Iafrate, A.J., Feuk, L., Rivera, M.N., Listewnik, M.L., Donahoe, P.K., Qi, Y., Scherer, S.W., and Lee, C. (2004). Detection of large-scale variation in the human genome. Nat. Genet. *36*, 949–951.

Ishkanian, A.S., Malloff, C.A., Watson, S. K., DeLeeuw, R.J., Chi, B., Coe, B.P., Snijders, A., Albertson, D.G., Pinkel, D., Marra, M.A., *et al.* (2004). A tiling resolution DNA microarray with complete coverage of the human genome. Nat. Genet. *36*, 299–303.

Iyer, V.R., Horak, C.E., Scafe, C.S., Botstein, D., Snyder, M., and Brown, P.O. (2001). Genomic binding sites of the yeasT-cell-cycle transcription factors SBF and MBF. Nature *409*, 533–538.

Jansen, R C., and Nap, J. P. (2001). Genetical genomics: the added value from segregation. Trends Genet. *17*, 388–391.

Jeon, Y., Bekiranov, S., Karnani, N., Kapranov, P., Ghosh, S., MacAlpine, D., Lee, C., Hwang, D. S., Gingeras, T.R., and Dutta, A. (2005). Temporal profile of replication of human chromosomes. Proc. Natl. Acad. Sci. USA *102*, 6419–6424.

Johnson, J.M., Castle, J., Garrett-Engele, P., Kan, Z., Loerch, P. M., Armour, C. D., Santos, R., Schadt, E. E., Stoughton, R., and Shoemaker, D.D. (2003). Genome-wide survey of human alternative pre-mRNA splicing with exon junction microarrays. Science *302*, 2141–2144.

Johnson, J.M., Edwards, S., Shoemaker, D., and Schadt, E.E. (2005). Dark matter in the genome: evidence of widespread transcription detected by microarray tiling experiments. Trends Genet. *21*, 93–102.

Kampa, D., Cheng, J., Kapranov, P., Yamanaka, M., Brubaker, S., Cawley, S., Drenkow, J., Piccolboni, A., Bekiranov, S., Helt, G., *et al.* (2004). Novel RNAs identified from an in-depth analysis of the transcriptome of human chromosomes 21 and 22. Genome Res. *14*, 331–342.

Kapranov, P., Cawley, S.E., Drenkow, J., Bekiranov, S., Strausberg, R.L., Fodor, S.P., and Gingeras, T.R. (2002). Large-scale transcriptional activity in chromosomes 21 and 22. Science *296*, 916–919.

Kapranov, P., Drenkow, J., Cheng, J., Long, J., Helt, G., Dike, S., and Gingeras, T.R. (2005). Examples of the complex architecture of the human transcriptome revealed by RACE and high-density tiling arrays. Genome Res. *15*, 987–997.

Kennedy, G. C., Matsuzaki, H., Dong, S., Liu, W. M., Huang, J., Liu, G., Su, X., Cao, M., Chen, W., Zhang, J., *et al.* (2003). Large-scale genotyping of complex DNA. Nat. Biotechnol. *21*, 1233–1237.

Keshet, I., Schlesinger, Y., Farkash, S., Rand, E., Hecht, M., Segal, E., Pikarski, E., Young, R.A., Niveleau, A., Cedar, H., and Simon, I. (2006). Evidence for an instructive mechanism of de novo methylation in cancer cells. Nat. Genet. *38*, 149–153.

Kostrzynska, M., and Bachand, A. (2006). Application of DNA microarray technology for detection, identification, and characterization of food-borne pathogens. Can. J. Microbiol. *52*, 1–8.

Krek, A., Grun, D., Poy, M.N., Wolf, R., Rosenberg, L., Epstein, E.J., MacMenamin, P., da Piedade, I., Gunsalus, K.C., Stoffel, M., and Rajewsky, N. (2005). Combinatorial microRNA target predictions. Nat. Genet. *37*, 495–500.

Lall, S., Grun, D., Krek, A., Chen, K., Wang, Y.L., Dewey, C.N., Sood, P., Colombo, T., Bray, N., Macmenamin, P., *et al.* (2006). A genome-wide map of conserved microRNA targets in C. elegans. Curr. Biol. *16*, 460–471.

Lau, N.C., Seto, A.G., Kim, J., Kuramochi-Miyagawa, S., Nakano, T., Bartel, D. P., and Kingston, R. E. (2006). Characterization of the piRNA Complex from Rat Testes. Science.

Leu, Y.W., Yan, P.S., Fan, M., Jin, V.X., Liu, J.C., Curran, E.M., Welshons, W.V., Wei, S. H., Davuluri, R.V., Plass, C., *et al.* (2004). Loss of estrogen receptor signaling triggers epigenetic silencing of downstream targets in breast cancer. Cancer Res. *64*, 8184–8192.

Li, C., Kato, M., Shiue, L., Shively, J. E., Ares, M., Jr., and Lin, R. J. (2006). Cell type and culture condition-dependent alternative splicing in human breast cancer cells revealed by splicing-sensitive microarrays. Cancer Res. *66*, 1990–1999.

Li, Z., Van Calcar, S., Qu, C., Cavenee, W.K., Zhang, M.Q., and Ren, B. (2003). A global transcriptional regulatory role for c-Myc in Burkitt's lymphoma cells. Proc. Natl. Acad. Sci. USA *100*, 8164–8169.

Lieb, J.D., Liu, X., Botstein, D., and Brown, P.O. (2001). Promoter-specific binding of Rap1 revealed by genome-wide maps of protein-DNA association. Nat. Genet. *28*, 327–334.

Lim, L.P., Lau, N.C., Garrett-Engele, P., Grimson, A., Schelter, J.M., Castle, J., Bartel, D.P., Linsley, P.S., and Johnson, J.M. (2005). Microarray analysis shows that some microRNAs downregulate large numbers of target mRNAs. Nature *433*, 769–773.

Lin, B., Wang, Z., Vora, G.J., Thornton, J.A., Schnur, J.M., Thach, D.C., Blaney, K.M., Ligler, A.G., Malanoski, A.P., Santiago, J., *et al.* (2006). Broad-spectrum respiratory tract pathogen identification using resequencing DNA microarrays. Genome Res. *16*, 527–535.

Liu, C.G., Calin, G.A., Meloon, B., Gamliel, N., Sevignani, C., Ferracin, M., Dumitru, C. D., Shimizu, M., Zupo, S., Dono, M., *et al.* (2004). An oligonucleotide microchip for genome-wide microRNA profiling in human and mouse tissues. Proc. Natl. Acad. Sci. USA *101*, 9740–9744.

Lodes, M.J., Suciu, D., Elliott, M., Stover, A.G., Ross, M., Caraballo, M., Dix, K., Crye, J., Webby, R J., Lyon, W.J., *et al.* (2006). Use of semiconductor-based oligonucleotide microarrays for influenza a virus subtype identification and sequencing. J. Clin .Microbiol. *44*, 1209–1218.

Lu, J., Getz, G., Miska, E. A., Alvarez-Saavedra, E., Lamb, J., Peck, D., Sweet-Cordero, A., Ebert, B.L., Mak, R.H., Ferrando, A.A., *et al.* (2005). MicroRNA expression profiles classify human cancers. Nature *435*, 834–838.

Lucito, R., Healy, J., Alexander, J., Reiner, A., Esposito, D., Chi, M., Rodgers, L., Brady, A., Sebat, J., Troge, J., *et al.* (2003). Representational oligonucleotide microarray analysis: a high-resolution method to detect genome copy number variation. Genome Res. *13*, 2291–2305.

Lum, P.Y., Chen, Y., Zhu, J., Lamb, J., Melmed, S., Wang, S., Drake, T.A., Lusis, A.J., and Schadt, E.E. (2006). Elucidating the murine brain transcriptional network in a segregating mouse population to identify core functional modules for obesity and diabetes. J. Neurochem.

Mantripragada, K.K., Tapia-Paez, I., Blennow, E., Nilsson, P., Wedell, A., and Dumanski, J.P. (2004). DNA copy-number analysis of the 22q11 deletion-syndrome region using array-CGH with genomic and PCR-based targets. Int. J. Mol. Med. *13*, 273–279.

Martinez-Climent, J. A., Alizadeh, A. A., Segraves, R., Blesa, D., Rubio-Moscardo, F., Albertson, D.G., Garcia-Conde, J., Dyer, M. J., Levy, R., Pinkel, D., and Lossos, I.S. (2003). Transformation of follicular lymphoma to diffuse large cell lymphoma is associated with a heterogeneous set of DNA copy number and gene expression alterations. Blood *101*, 3109–3117.

Massion, P.P., Kuo, W.L., Stokoe, D., Olshen, A.B., Treseler, P.A., Chin, K., Chen, C., Polikoff, D., Jain, A.N., Pinkel, D., *et al.* (2002). Genomic copy number analysis of non-small cell lung cancer using array comparative genomic hybridization: implications of the phosphatidylinositol 3-kinase pathway. Cancer Res. *62*, 3636–3640.

Matsuzaki, H., Dong, S., Loi, H., Di, X., Liu, G., Hubbell, E., Law, J., Berntsen, T., Chadha, M., Hui, H., *et al.* (2004a). Genotyping over 100,000 SNPs on a pair of oligonucleotide arrays. Nat. Methods *1*, 109–111.

Matsuzaki, H., Loi, H., Dong, S., Tsai, Y.Y., Fang, J., Law, J., Di, X., Liu, W. M., Yang, G., Liu, G., *et al.* (2004b). Parallel genotyping of over 10,000 SNPs using a one-primer assay on a high-density oligonucleotide array. Genome Res. *14*, 414–425.

Mattick, J.S., and Makunin, I.V. (2006). Non-coding RNA. Hum. Mol. Genet. *15 Spec. No. 1*, R17–29.

Mattie, M.D., Benz, C.C., Bowers, J., Sensinger, K., Wong, L., Scott, G.K., Fedele, V., Ginzinger, D.G., Getts, R.C., and Haqq, C.M. (2006). Optimized high-throughput microRNA expression profiling provides novel biomarker assessment of clinical prostate and breast cancer biopsies. Mol. Cancer *5*, 24.

Mestre, C., Rubio-Moscardo, F., Rosenwald, A., Climent, J., Dyer, M. J., Staudt, L., Pinkel, D., Siebert, R., and Martinez-Climent, J.A. (2005). Homozygous deletion of SOCS1 in primary mediastinal B-cell lymphoma detected by CGH to BAC microarrays. Leukemia *19*, 1082–1084.

Meyers, B.C., Souret, F.F., Lu, C., and Green, P.J. (2006). Sweating the small stuff: microRNA discovery in plants. Curr. Opin. Biotechnol. *17*, 139–146.

Miller, R.D., Phillips, M.S., Jo, I., Donaldson, M.A., Studebaker, J.F., Addleman, N., Alfisi, S.V., Ankener, W.M., Bhatti, H.A., Callahan, C.E., *et al.* (2005). High-density single-nucleotide polymorphism maps of the human genome. Genomics *86*, 117–126.

Miska, E.A., Alvarez-Saavedra, E., Townsend, M., Yoshii, A., Sestan, N., Rakic, P., Constantine-Paton, M., and Horvitz, H.R. (2004). Microarray analysis of microRNA expression in the developing mammalian brain. Genome Biol. *5*, R68.

Modrek, B., and Lee, C. (2002). A genomic view of alternative splicing. Nat. Genet. *30*, 13–19.

Monticelli, S., Ansel, K. M., Xiao, C., Socci, N. D., Krichevsky, A. M., Thai, T. H., Rajewsky, N., Marks, D.S., Sander, C., Rajewsky, K., *et al.* (2005). MicroRNA profiling of the murine hematopoietic system. Genome Biol. *6*, R71.

Nagao, K., Togawa, N., Fujii, K., Uchikawa, H., Kohno, Y., Yamada, M., and Miyashita, T. (2005). Detecting tissue-specific alternative splicing and disease-associated aberrant splicing of the PTCH gene with exon junction microarrays. Hum. Mol. Genet. *14*, 3379–3388.

Nuwaysir, E.F., Huang, W., Albert, T.J., Singh, J., Nuwaysir, K., Pitas, A., Richmond, T., Gorski, T., Berg, J.P., Ballin, J., *et al.* (2002). Gene expression analysis using oligonucleotide arrays produced by maskless photolithography. Genome Res. *12*, 1749–1755.

Orian, A. (2006). Chromatin profiling, DamID and the emerging landscape of gene expression. Curr. Opin. Genet. Dev. *16*, 157–164.

Orian, A., van Steensel, B., Delrow, J., Bussemaker, H.J., Li, L., Sawado, T., Williams, E., Loo, L. W., Cowley, S.M., Yost, C., *et al.* (2003). Genomic binding by the Drosophila Myc, Max, Mad/Mnt transcription factor network. Genes Dev. *17*, 1101–1114.

Ouahchi, K., Lindeman, N., and Lee, C. (2006). Copy number variants and pharmacogenomics. Pharmacogenomics *7*, 25–29.

Pan, Q., Saltzman, A.L., Kim, Y.K., Misquitta, C., Shai, O., Maquat, L. E., Frey, B.J., and Blencowe, B.J. (2006). Quantitative microarray profiling provides evidence against widespread coupling of alternative splicing with nonsense-mediated mRNA decay to control gene expression. Genes Dev. *20*, 153–158.

Pan, Q., Shai, O., Misquitta, C., Zhang, W., Saltzman, A. L., Mohammad, N., Babak, T., Siu, H., Hughes, T.R., Morris, Q.D., *et al.* (2004). Revealing global regulatory features of mammalian alternative splicing using a quantitative microarray platform. Mol. Cell *16*, 929–941.

Peng, W.T., Robinson, M.D., Mnaimneh, S., Krogan, N.J., Cagney, G., Morris, Q., Davierwala, A. P., Grigull, J., Yang, X., Zhang, W., *et al.* (2003). A panoramic view of yeast noncoding RNA processing. Cell *113*, 919–933.

Perry, G.H., Tchinda, J., McGrath, S.D., Zhang, J., Picker, S.R., Caceres, A.M., Iafrate, A. J., Tyler-Smith, C., Scherer, S. W., Eichler, E E., *et al.* (2006). Hotspots for copy number variation in chimpanzees and humans. Proc. Natl. Acad. Sci. USA *103*, 8006–8011.

Pinkel, D., and Albertson, D.G. (2005). Array comparative genomic hybridization and its applications in cancer. Nat. Genet. *37*, Suppl., S11–17.

Pokholok, D.K., Harbison, C.T., Levine, S., Cole, M., Hannett, N.M., Lee, T.I., Bell, G.W., Walker, K., Rolfe, P.A., Herbolsheimer, E., *et al.* (2005). Genome-wide map of nucleosome acetylation and methylation in yeast. Cell *122*, 517–527.

Rajewsky, N. (2006). microRNA target predictions in animals. Nat. Genet. *38*, Suppl. *1*, S8-S13.

Raskin, D.M., Seshadri, R., Pukatzki, S.U., and Mekalanos, J.J. (2006). Bacterial genomics and pathogen evolution. Cell *124*, 703–714.

Ravasi, T., Suzuki, H., Pang, K.C., Katayama, S., Furuno, M., Okunishi, R., Fukuda, S., Ru, K., Frith, M. C., Gongora, M.M., *et al.* (2006). Experimental validation of the regulated expression of large numbers of non-coding RNAs from the mouse genome. Genome Res. *16*, 11–19.

Ren, B., Robert, F., Wyrick, J.J., Aparicio, O., Jennings, E.G., Simon, I., Zeitlinger, J., Schreiber, J., Hannett, N., Kanin, E., *et al.* (2000). Genome-wide location and function of DNA binding proteins. Science *290*, 2306–2309.

Rhodes, D.R., and Chinnaiyan, A.M. (2005). Integrative analysis of the cancer transcriptome. Nat. Genet. *37 Suppl*, S31–37.

Rinn, J.L., Euskirchen, G., Bertone, P., Martone, R., Luscombe, N. M., Hartman, S., Harrison, P. M., Nelson, F. K., Miller, P., Gerstein, M., *et al.* (2003). The transcriptional activity of human Chromosome 22. Genes Dev. *17*, 529–540.

Rodriguez, B.A., and Huang, T.H. (2005). Tilling the chromatin landscape: emerging methods for the discovery and profiling of protein-DNA interactions. Biochem. Cell Biol. *83*, 525–534.

Ross-Macdonald, P., Coelho, P.S., Roemer, T., Agarwal, S., Kumar, A., Jansen, R., Cheung, K.H., Sheehan, A., Symoniatis, D., Umansky, L., *et al.* (1999). Large-scale analysis of the yeast genome by transposon tagging and gene disruption. Nature *402*, 413–418.

Saha, A., Wittmeyer, J., and Cairns, B.R. (2006). Chromatin remodelling: the industrial revolution of DNA around histones. Nat. Rev. Mol. Cell. Biol. *7*, 437–447.

Schadt, E.E. (2005). Exploiting naturally occurring DNA variation and molecular profiling data to dissect disease and drug response traits. Curr. Opin. Biotechnol. *16*, 647–654.

Schadt, E.E., Edwards, S. W., GuhaThakurta, D., Holder, D., Ying, L., Svetnik, V., Leonardson, A., Hart, K.W., Russell, A., Li, G., *et al.* (2004). A comprehensive transcript index of the human genome generated using microarrays and computational approaches. Genome Biol. *5*, R73.

Schadt, E.E., Lamb, J., Yang, X., Zhu, J., Edwards, S., Guhathakurta, D., Sieberts, S. K., Monks, S., Reitman, M., Zhang, C., *et al.* (2005). An integrative genomics approach to infer causal associations between gene expression and disease. Nat. Genet. *37*, 710–717.

Schadt, E.E., Monks, S.A., Drake, T.A., Lusis, A.J., Che, N., Colinayo, V., Ruff, T. G., Milligan, S.B., Lamb, J.R., Cavet, G., *et al.* (2003). Genetics of gene expression surveyed in maize, mouse and man. Nature *422*, 297–302.

Schubeler, D., MacAlpine, D.M., Scalzo, D., Wirbelauer, C., Kooperberg, C., van Leeuwen, F., Gottschling, D.E., O'Neill, L.P., Turner, B.M., Delrow, J., *et al.* (2004). The histone modification pattern of active genes revealed through genome-wide chromatin analysis of a higher eukaryote. Genes Dev. *18*, 1263–1271.

Schumacher, A., Kapranov, P., Kaminsky, Z., Flanagan, J., Assadzadeh, A., Yau, P., Virtanen, C., Winegarden, N., Cheng, J., Gingeras, T., and Petronis, A. (2006). Microarray-based DNA methylation profiling: technology and applications. Nucleic Acids Res. *34*, 528–542.

Schwartz, Y.B., Kahn, T.G., Nix, D.A., Li, X.Y., Bourgon, R., Biggin, M., and Pirrotta, V. (2006). Genome-wide analysis of Polycomb targets in Drosophila melanogaster. Nat. Genet. *38*, 700–705.

Scott, G.K., Mattie, M.D., Berger, C. E., Benz, S.C., and Benz, C.C. (2006). Rapid alteration of microRNA levels by histone deacetylase inhibition. Cancer Res. *66*, 1277–1281.

Sebat, J., Lakshmi, B., Troge, J., Alexander, J., Young, J., Lundin, P., Maner, S., Massa, H., Walker, M., Chi, M., *et al.* (2004). Large-scale copy number polymorphism in the human genome. Science *305*, 525–528.

Segal, E., Friedman, N., Kaminski, N., Regev, A., and Koller, D. (2005). From signatures to models: understanding cancer using microarrays. Nat. Genet. *37* Suppl., S38–45.

Segal, E., Friedman, N., Koller, D., and Regev, A. (2004). A module map showing conditional activity of expression modules in cancer. Nat. Genet. *36*, 1090–1098.

Selinger, D.W., Cheung, K.J., Mei, R., Johansson, E.M., Richmond, C.S., Blattner, F.R., Lockhart, D.J., and Church, G.M. (2000). RNA expression analysis using a 30 base pair resolution Escherichia coli genome array. Nat. Biotechnol. *18*, 1262–1268.

Selzer, R.R., Richmond, T.A., Pofahl, N.J., Green, R.D., Eis, P.S., Nair, P., Brothman, A.R., and Stallings, R.L. (2005). Analysis of chromosome breakpoints in neuroblastoma at sub-kilobase resolution using fine-tiling oligonucleotide array CGH. Genes Chromosomes Cancer *44*, 305–319.

Semon, M., and Duret, L. (2004). Evidence that functional transcription units cover at least half of the human genome. Trends Genet. *20*, 229–232.

Sharp, A.J., Locke, D.P., McGrath, S.D., Cheng, Z., Bailey, J.A., Vallente, R.U., Pertz, L.M., Clark, R.A., Schwartz, S., Segraves, R., *et al.* (2005). Segmental duplications and copy-number variation in the human genome. Am. J. Hum. Genet. *77*, 78–88.

Shoemaker, D.D., Schadt, E.E., Armour, C.D., He, Y.D., Garrett-Engele, P., McDonagh, P.D., Loerch, P.M., Leonardson, A., Lum, P. Y., Cavet, G., *et al.* (2001). Experimental annotation of the human genome using microarray technology. Nature *409*, 922–927.

Slater, H. R., Bailey, D.K., Ren, H., Cao, M., Bell, K., Nasioulas, S., Henke, R., Choo, K. H., and Kennedy, G. C. (2005). High-resolution identification of chromosomal abnormalities using oligonucleotide arrays containing 116,204 SNPs. Am. J. Hum. Genet. *77*, 709–726.

Song, S., Cooperman, J., Letting, D.L., Blobel, G.A., and Choi, J.K. (2004). Identification of cyclin D3 as a direct target of E2A using DamID. Mol. Cell Biol. *24*, 8790–8802.

Sood, P., Krek, A., Zavolan, M., Macino, G., and Rajewsky, N. (2006). Cell-type-specific signatures of microRNAs on target mRNA expression. Proc. Natl. Acad. Sci. USA *103*, 2746–2751.

Srinivasan, K., Shiue, L., Hayes, J.D., Centers, R., Fitzwater, S., Loewen, R., Edmondson, L. R., Bryant, J., Smith, M., Rommelfanger, C., *et al.* (2005). Detection and measurement of alternative splicing using splicing-sensitive microarrays. Methods *37*, 345–359.

Stolc, V., Gauhar, Z., Mason, C., Halasz, G., van Batenburg, M. F., Rifkin, S. A., Hua, S., Herreman, T., Tongprasit, W., Barbano, P. E., *et al.* (2004). A gene expression map for the euchromatic genome of Drosophila melanogaster. Science *306*, 655–660.

Storz, G. (2002). An expanding universe of noncoding RNAs. Science *296*, 1260–1263.

Syvanen, A.C. (2005). Toward genome-wide SNP genotyping. Nat. Genet. *37 Suppl*, S5–10.

Tjaden, B., Goodwin, S.S., Opdyke, J.A., Guillier, M., Fu, D.X., Gottesman, S., and Storz, G. (2006). Target prediction for small, noncoding RNAs in bacteria. Nucleic Acids Res. *34*, 2791–2802.

Tuzun, E., Sharp, A.J., Bailey, J.A., Kaul, R., Morrison, V. A., Pertz, L.M., Haugen, E., Hayden, H., Albertson, D., Pinkel, D., *et al.* (2005). Fine-scale structural variation of the human genome. Nat. Genet. *37*, 727–732.

van den Ijssel, P., Tijssen, M., Chin, S. F., Eijk, P., Carvalho, B., Hopmans, E., Holstege, H., Bangarusamy, D.K., Jonkers, J., Meijer, G.A., *et al.* (2005). Human and mouse oligonucleotide-based array CGH. Nucleic Acids Res. 33, e192.

van Steensel, B. (2005). Mapping of genetic and epigenetic regulatory networks using microarrays. Nat. Genet. 37 Suppl., S18–24.

van Steensel, B., Delrow, J., and Henikoff, S. (2001). Chromatin profiling using targeted DNA adenine methyltransferase. Nat. Genet. *27*, 304–308.

van Steensel, B., and Henikoff, S. (2000). Identification of in vivo DNA targets of chromatin proteins using tethered dam methyltransferase. Nat. Biotechnol. *18*, 424–428.

van Steensel, B., and Henikoff, S. (2003). Epigenomic profiling using microarrays. Biotechniques *35*, 346–350, 352–344, 356–347.

van't Veer, L.J., Dai, H., van de Vijver, M.J., He, Y.D., Hart, A.A., Mao, M., Peterse, H.L., van der Kooy, K., Marton, M. J., Witteveen, A. T., *et al.* (2002). Gene expression profiling predicts clinical outcome of breast cancer. Nature *415*, 530–536.

Vissers, L.E., Veltman, J. A., van Kessel, A. G., and Brunner, H. G. (2005). Identification of disease genes by whole genome CGH arrays. Hum. Mol. Genet. *14* Spec. No. *2*, R215–223.

Volinia, S., Calin, G. A., Liu, C. G., Ambs, S., Cimmino, A., Petrocca, F., Visone, R., Iorio, M., Roldo, C., Ferracin, M., *et al.* (2006). A microRNA expression signature of human solid tumors defines cancer gene targets. Proc. Natl. Acad. Sci. USA *103*, 2257–2261.

Wang, D., Coscoy, L., Zylberberg, M., Avila, P.C., Boushey, H.A., Ganem, D., and DeRisi, J.L. (2002). Microarray-based detection and genotyping of viral pathogens. Proc. Natl. Acad. Sci. USA *99*, 15687–15692.

Wang, X. (2006). Systematic identification of microRNA functions by combining target prediction and expression profiling. Nucleic Acids Res. *34*, 1646–1652.

Warrington, J.A., Shah, N.A., Chen, X., Janis, M., Liu, C., Kondapalli, S., Reyes, V., Savage, M. P., Zhang, Z., Watts, R., *et al.* (2002). New developments in high-throughput resequencing and variation detection using high density microarrays. Hum. Mutat. *19*, 402–409.

Washietl, S., Hofacker, I. L., Lukasser, M., Huttenhofer, A., and Stadler, P. F. (2005a). Mapping of conserved RNA secondary structures predicts thousands of functional noncoding RNAs in the human genome. Nat. Biotechnol. *23*, 1383–1390.

Washietl, S., Hofacker, I.L., and Stadler, P.F. (2005b). Fast and reliable prediction of noncoding RNAs. Proc. Natl. Acad. Sci. USA *102*, 2454–2459.

Weber, M., Davies, J. J., Wittig, D., Oakeley, E. J., Haase, M., Lam, W. L., and Schubeler, D. (2005). Chromosome-wide and promoter-specific analyses identify sites of differential DNA methylation in normal and transformed human cells. Nat. Genet. *37*, 853–862.

Wei, S.H., Chen, C.M., Strathdee, G., Harnsomburana, J., Shyu, C.R., Rahmatpanah, F., Shi, H., Ng, S.W., Yan, P.S., Nephew, K.P., *et al.* (2002). Methylation microarray analysis of late-stage ovarian carcinomas distinguishes progression-free survival in patients and identifies candidate epigenetic markers. Clin. Cancer Res. *8*, 2246–2252.

Weil, M.R., Widlak, P., Minna, J.D., and Garner, H.R. (2004). Global survey of chromatin accessibility using DNA microarrays. Genome Res. *14*, 1374–1381.

Whitfield, M.L., George, L.K., Grant, G.D., and Perou, C.M. (2006). Common markers of proliferation. Nat. Rev. Cancer *6*, 99–106.

Wilson, I.M., Davies, J.J., Weber, M., Brown, C.J., Alvarez, C.E., MacAulay, C., Schubeler, D., and Lam, W.L. (2006). Epigenomics: mapping the methylome. Cell Cycle *5*, 155–158.

Yamada, K., Lim, J., Dale, J.M., Chen, H., Shinn, P., Palm, C.J., Southwick, A. M., Wu, H. C., Kim, C., Nguyen, M., *et al.* (2003). Empirical analysis of transcriptional activity in the Arabidopsis genome. Science *302*, 842–846.

Yan, P.S., Chen, C.M., Shi, H., Rahmatpanah, F., Wei, S.H., Caldwell, C.W., and Huang, T.H. (2001). Dissecting complex epigenetic alterations in breast cancer using CpG island microarrays. Cancer Res. *61*, 8375–8380.

Yan, P.S., Wei, S.H., and Huang, T.H. (2004). Methylation-specific oligonucleotide microarray. Methods Mol. Biol. *287*, 251–260.
Yanaihara, N., Caplen, N., Bowman, E., Seike, M., Kumamoto, K., Yi, M., Stephens, R. M., Okamoto, A., Yokota, J., Tanaka, T., *et al.* (2006). Unique microRNA molecular profiles in lung cancer diagnosis and prognosis. Cancer Cell *9*, 189–198.
Ylstra, B., van den Ijssel, P., Carvalho, B., Brakenhoff, R. H., and Meijer, G. A. (2006). BAC to the future! or oligonucleotides: a perspective for micro array comparative genomic hybridization (array CGH). Nucleic Acids Res. *34*, 445–450.
Zhang, C., Li, H. R., Fan, J. B., Wang-Rodriguez, J., Downs, T., Fu, X. D., and Zhang, M. Q. (2006). Profiling alternatively spliced mRNA isoforms for prostate cancer classification. BMC Bioinformatics *7*, 202.
Zhang, Y., Zhang, Z., Ling, L., Shi, B., and Chen, R. (2004). Conservation analysis of small RNA genes in *Escherichia coli*. Bioinformatics *20*, 599–603.
Zhao, X., Li, C., Paez, J. G., Chin, K., Janne, P. A., Chen, T. H., Girard, L., Minna, J., Christiani, D., Leo, C., *et al.* (2004). An integrated view of copy number and allelic alterations in the cancer genome using single nucleotide polymorphism arrays. Cancer Res. *64*, 3060–3071.
Zhou, D., Han, Y., Dai, E., Song, Y., Pei, D., Zhai, J., Du, Z., Wang, J., Guo, Z., and Yang, R. (2004). Defining the genome content of live plague vaccines by use of whole-genome DNA microarray. Vaccine *22*, 3367–3374.

Index

A

ABCB1 80–81
ABCB3 80
ABCB5 80–81
Absorption 71,80
Acetylation 154
Acute lymphoblastic leukemia (ALL) 11,105–108
Acute myeloid leukemia (AML) 11,33–34,105–108
Adenine transferase *see* DamID
ADME 71
ADME/Tox 71,83
ADME-AP 83
Administration, IL12 123
Affymetrix 5,38,124,129–130,153
- absent probeset 134
- normalization 133
- probe set intensity 132

Agilent 6
Alkaline phosphatase *see* ALT
Allogeneic 123
ALT 74,76
Alternative splicing 144–146
Amplification 7,15
Animal virus 157
Annotation 137
ANOVA 56,102,136
Antibody 153
Apoptosis 81
Arabidopsis 147
ARACNE (algorithm for the reconstruction of accurate cellular network) 41,64
Area under the curve (AUC) 84
Aromatic polycycles 75
Array-CGH *see* CGH
Asparagine synthetase 37
Assay 73
Atlas, examples 90
AUC *see* Area under the curve
Average linkage *see* UPGMA
Azacytidine 80

B

BAC clone 150
Backward propagation (BP) 110
Bacteria 156,157
- evolution 157
- phylogenomics 157

Bacteria, virulence 157
Baker's antifol 81
Bayes' rule 44
Bayesian classifier 79
Bayesian Dirichlet equivalence (BDe) 44
Bayesian information criteria (BIC) 44
B-cell lymphoma 11,102
B-cells 64,125
Biomarkers 78,93–95
- signature 34

Bisulfite 151
BMD *see* Bone mineral density
Bone 94–95
- mineral density 94

Brain, subsections 97
BRCA1 14
Breast cancer 12,121–130
Bromochloroacetic acid 76
Budding yeast 156
Burkitt's lymphoma 153

C

Cancer 59,121–124,143,149
- animal model 122–123
- carcinogenesis 121–124
- gene modules 59
- microRNA 149
- transcriptome 143
- vaccines 121–124

Carcinogenesis 81,121–124
Carcinogenicity 76
Carcinogens 76
Carcinoma, mammary gland 122
cDNA 7
CEBS 83
Cell cycle 66,81
Cell lines 95
Cellular toxicity 73
Centroid linkage *see* UPGMC
Centroids 19
CFTR 13
CGH 150,155
CGI 153
Channel 71
Chemical Effects in Biological Systems *see* CEBS
Chemokine 124
Chemoresistance 81
ChIP-on-chip 13,151–154
Chromatin immunoprecipitation (ChIP) 13,64,65,152–153
Chromatin structure 150,154
Chromatin 154
- condensation state 154
- modification 154
- structure 154

Chromosomes 21 and 22 153
Class comparison 102
Class discovery 102
Class prediction 104–111
Classification 17,79,102
- model 79
- problem 79
- tree (CT) 38

Classifier 31–32,37–39
Clinical follow-up 11
Clinical phase 30,71
Clinical samples 157
Clustering 17,18,32
Clustering, hierarchical 126
C-Myc 153
Co-expression 99
Collaborative Cross 156
Comparative genome hybridization *see* CGH
Compendium, toxicity prediction 83
Complete linkage 18
Complex trait 156
Component, causative 156
Component, reactive 156
Compound 71,73,78
- classification 78
- development 73

Confocal microscopy 127
Contamination, tissue 93,95
Copy-number 150
Copy-number variation 156
Coronavirus 156
Correlation, gene expression 95,98
Cortical bone 94–95
Covalent modification 154
Covariance 22
CpG 151
CpG-island *see* CGI
CRADD 128–129
cRNA 7
Cross-linking 152–153
Cross-Species 99
Cross-validation 105
CuraGen 83
CyberT 136
CYP450 *see* cytochrome P450
Cytochrome P450 73,76,93–94
Cytokines 124
Cytotoxic mechanism 81

D

DAG 64
DamID 155
Data analysis 15–16,129–137
- annotation 137
- artifacts 130–131
- filtering 134
- normalization 15,133–135
- replicates 130–132
- sample homogeneity 132
- statistical validation 135

Data 16,22,38,43,79,81–82,134
- dimensionality 16
- filtering 134
- integration 81–82
- processing inequality (DPI) 43
- reduction 79
- test training 22,38

Database 23,33,59–60,83,108,137
- ArrayExpress 83
- CEBS 83
- CIBEX 83
- GENMAPP 59
- GEO 83
- KEGG 59–60
- membrane transporter 83
- metabolism 83
- metabolite 83
- MGED (Microarray Gene Expression Data) 23
- MIPS (Munich Information Center for Protein Sequences) 108
- MYGD (MIPS Yeast Genome Database) 33
- nuclear receptor database 83
- PharmaGKB 83

Stanford Microarray Database (SMD) 23
toxicogenomics 82–83
Unigene 137
Dendogram 19
Diabetes 91,156
Diagnostic marker 71
Diagnostics 11,14
Diaphysis 95
Difference, organ 97
Difference, Species 97
Differential expression 127
Dimethylation 154
Directed acyclic graph (DAG) 45,56
Disease 8
Disease, development 8
Distance similarity 19–20,37,61,115
euclidean distance 20,115
pearson's correlation 22,37
Distribution 71
DNA computing 8
DNA damage 41,49,66,81
DNA repair 81
DNA vaccination, intramuscular 123
DNA-reactive compound 81
Dorsal root ganglion 90
Dose 74
Drosophila melanogaster 147,153
Drug 8,10–14,23,29–34,37–40,49–50,71–73,80,84,89–97,156
adverse effects 71
antidepressants (ADs) 38–39
antimicrobial 49
antimycotic 33
antipsycotics (APs) 38
development 73
development 30
discovery 73,89–97
discovery 29–31,39,40,50
efficacy 31,34,37
inhibitor 8
mechanism of action (MOA) 10,30–31,34–35
metabolism 84
resistance 80
response 10–14,23,31,72,156
screening 10,12
signature 34
target 8
toxicity 12,31,37
treatment 10–11
network 49
Dynamical models 68

E

E2F1 97,99
ENCODE 153
End point 78
Enrichment, gene expression 93
Entropy 42
Enzyme 71
Epigenetics 152
Epigenomics 150,154
ERBB2 *see* HER2 121–130
ERG11 35
ERG2 35
Estrogen receptor 154
Euchromatin 154
Ewing family of tumors. 104
Excretion 71
Exon 144
Exosome 148
Experimental design 15–16
Extracellular compartment 128

F

False-negatives (FNs) 110
False-positives (FPs) 110,136
Fingerprint 10,23
FK506 35
Fluorescent dye 2,5,6,15
Cy5-Cy3 7
efficiency 7–8,15
non-linearity 15
saturation 15
Fluorophor 6–7
Food 157
F-test 79
Functional Class Scoring (FCS) 59
Funnel, testing 73–74

G

Galnt3 128–129
GCRMA 132
GE-HTS (gene expression based high throughput screening) 33–34
Geldanamycin 81
Gene expression based high throughput screening *see* GE-HTS
Gene Expression Omnibus (GEO) 23
Gene expression signature 33,36
Gene Logic 83
Gene Ontology (GO) 49,56–59,128–129,136
enrichment 128–129
Gene promoters 40,151–153
methylation 151
Gene Set Enrichment Analysis (GSEA) 58
Gene sets analysis 58
GeneChip 130
Gene 7,9,12,17,31,63,97–99,127
amplification 12

antibody-related 127
classification 17,31
down-regulated 7
gene set 9
genes module 9,63
humoral response 127
orthologous 97–98
target 97–99
up-regulated 7
Genome 12,71–72,80,146–148,150,155–157,
bacteria 148,156–157
eukaryotic 148
gene-coding region 146–147
intergenic regions 148
methylation 149–150,155
pharmacogenomics 72
proteomics 72
structural variation 156
toxicogenomics 71
transcriptionally functional units 147
transportome 80
variation 155
virus 156
Genomic DNA 12
Genomic tiling 146–148
Genomics 11,156
integrative 156
Genotoxic mechanism 81
Genotoxicity 76,81
Genotyping 2,5,13,156
Genus, viral 156
GO *see* Gene ontology
GO terms 56–59,62
Graph theory 39
Greedy-hill climbing 43
Growth inhibition (GI) 37–38
Guilt-by-association 18

H

H-2q 123
Heatmap 9,19,65,108
Hepatotoxicants 78
Hepatotoxicity 78
hepatotoxin 80
HER2 121–127,130
Heterozygosity 155
Hierarchical clustering 18–20,36,37
High throughput screening *see* HTS
High-density exon arrays 146
Hip 94–95
Histone 154
Histopathological changes 74
Hit identification 30,31,37,45
HTS (high throughput screening) 30
Hybridization 1,6–7,13–14
background 14
unspecific 14
Hypermethylation 153
Hyperplasia 122
Hypertension 156

I

Iconix 83
IFNG *see* Interferon gamma
Ig A 127
Ig J 127
IGR *see* intergenic regions
IL12 123
IL-12 *see* Il12
Image segmentation 14
Immune response genes 125
Immunology, tumor 124
Immunoprecipitation 151,153
Immunostimulation 123
Immunotherapy 124
Inference 17
Inflammation 125
Influenza 157
Inhibitor 73
ink-jet 6
Inosine-glycodialdehyde 80
Insulin 91
Integrative genomics 156
Interferon gamma 123,125
Intergenic regions 148
Internalization 128
In-vitro transcription 15
Ionizing radiation 81
Islets of Langerhans 91,93

K

Kernel functions 107
k-means 20
k-nearest neighbor (kNN) 37,104–107,114–116
Kolmogorov-Smirnov 58

L

LAB-ON-A-CHIP 157
Labeling 7,15
Laser scanning 6,7
L-asparagine 37
LDA *see* Linear discriminant analysis
Lead identification 30,31,37,45
Lead optimization 30,31,37,45
Leave-one-out 37,38
Lesion 123
hyperplastic 123
metastatic 123
neoplastic 123
Leukemia cell line 37

Linear discriminant analysis
(LDA) 32,36,39
Liquid association 63
Live food 157
Liver 74
LNA 149
Loss function 32
Lymphoblastic leukemia 11

M

Machine learning 22
Mantel test 34
Marker, surrogate 74
Markov chain Montecarlo 43
MAS 132
Mass spectrometry 33
Mechanistic toxicogenomics 73
Meta-analysis 128
Metabolic profiling 77
Metabolism 71,84
prediction 84
Metabolites 77
Metabonomics 72,77
Metaphysis 95
Metastasis 123
Methylation 149–151,154,155
Methylome 150
MIAME 83
Microarray 1,6–7,13,15–16,124,130,144–
155
applications 144
customizable 146,149,153
double channel 6–7
exon probe 144–145
experiment 6–7
experimental design 15–16
genome-wide 146
genomic tiling 146–148
high-density 146,148
hybridization 1,6–7,13,130
intron probe 146
junction probe 145
LNA 149
methylation 149–151
MG-U74Av2 124
multi channel 6
quality 16
sample amplification 146
single channel 6
SNP 155
Microbial pathogens 13
Microbial strains 13
microRNA 149
3′ untranslated region 149
prognosis 149
Microrray, washing 130
mithocondrial DNA 14
MM *see* Probe, mismatch pair
MOA *see* Drug, mechanism of action
Model topology 43
Mode-of-action 45,48,76
by network identification (MNI) 45,48
Monkey 91,97–98
Montecarlo simulations 43
Mouse 122–124,156
BALB-neuT 122–124
genotyping 156
syngenic 122
trait 156
transgenic 122
mRNA 144–146
alternative splicing 144–146
exon 144
isoform 145
junction 145
Multilayer perceptrons (MLPs) 110
Multiple linear ridge regression 41
Multiple testing procedures 136
Muscle, heart 97
Muscle, skeletal 97
Mutations 68
MYC 43,64
Mycobacterium tuberculosis 13,36
Myofibril 97

N

NCI60 34,38
ncRNA *see* RNA, non-coding
Nearest shrunken centroids (NSCs) 103–
105,113
Nearest-neighbor (NN) 104
Network 9–10,23,30–31,39,41–45,61–
65,79,82,104,110,156
bayesian 43–44,63–65
functional 10
identification by regression (NIR) 41,45
interaction 156
neural 23,79,104,110
reconstruction 10,41,61,63
regulatory 10
reverse engineering 39,40
second-order analysis 63
signature 82
NEU *see* HER2
Neural synapse 156
Neuroblastoma (NB) 104
neuT 122
Nimblegen 153
NMR 77
Noise 14,79
reduction methods 79
source 14–15

non-Hodgkin lymphoma (NHL) 104
Nonsense-mediated mRNA decay 146
Normalization 15,133–135
  cyclic loess 134
  quantile 134
Nuclear magnetic resonance *see* NMR 77

O

Obesity 156
Omics 72
Oncogene 122
Opioid receptor agonists (OPs) 38
ORC1 66
Organ, mRNA abundance 89–99
Orthologous gene 97–98
Osteoporosis 94
Over-expression, HER2 122
Oxidoreductase 76

P

P53 14,63,81,153
Pain treatment 90–91
PAM (Prediction Analysis of Microarrays) 116
Pancreatic islets *see* Islets of Langerhans
Partial least square discriminant analysis 36
Partitioning methods 19
Pathophysiological stimuli 77
Pathway 9,33,35,55,58–60,81,127
  coherence 59–60
  druggable 33
  glycolysis 60
  HER2 degradation pathway 127
  HER2 signal transduction pathway 127
  oxidative phosphorylation pathway 58
  calcineurin signaling 35
Pattern recognition 77
PBPK 76,84
PCA *see* Principal Component Analysis
Permutation test 79
PGC1-alpha 59
Pharmaceutical research 8
Pharmacogenomics 11,72
Pharmacokinetics 12
Pharmacology 11
PharmaGKB 83
Phenotype 10,74–76
  pharmacological 74
  toxicological 74
  anchoring 74
Phylogenomics 157
Physiologically based pharmacokinetic modeling *see* PBPK
piRNA 148
Plague 157
Plasma level 80
PM *see* Probe, perfect match pair
polyA 146
Polycomb targets 154
Polymorphism 72
Post-translational modifications 60
Pre-classification 10
Preclinical phase 30
Preclinical testing 73–74
Predictive accuracy 107
Predictive model 78
Predictive toxicogenomics 78
Predictor variables 102
Premature termination codon 146
Pre-processing 17
Principal component analysis (PCA) 36–37,49,79,124–125
Probe 1–5,7–8,15,129–130
  amplification 5
  cloning 5
  hybridization 5
  in situ synthesis 2–5
  mismatch pair (MM) 5,8
  pairs 5
  PCR 5,7,15
  perfect match pair (PM) 5,8,129–130
  pre-synthesis 2–5
  probe set 5
  sequences 1
  spotting 5
Probe set intensity 130,132
Profiling, metabolic 77
Prognostics 11,14
Protein interaction network 82
Protein–DNA interaction 8,152–155
Proteome 2
Proteomics 2,72,76–77
PTCH 146
p-value 136

Q

QSAR 84
Quadratic programming QP 113
Quantitative structure–activity relationships *see* QSAR

R

Radiation, ionizing 81
Random forests (RF) 23,38–39,111,117
Rat 97–98
Rcn2 128–129
regulation motifs 62
Replicates 14,16,130–132
  biological 16
  experiments 14
  technical 16
Replication, temporal profile of 147
Response variables 102
Restriction enzyme 151,155

Reverse transcriptase 7,15
RFAM 148
Rhabdomyosarcoma (RMS) 104
r-HER2 *see* HER2
RMA *see* Robust Multi-Array analysis
RNA 81,148–149
  interference 81
  microRNA 149
  non-coding 148–149
  piRNA 148
  small 148–149
Rnf4 128–129
Robust Multi-Array analysis 124,132
R-squared coefficient 131
RT-PCR 33

S

SAM *see* Signficance Analysis of Microarrays
Sample 6,132,146
  amplification 146
  homogeneity 132
  reference 6
  test 6
Sampling over gene space (SOGS) 39
SAR (structure-activity relationship) 30
Sarcomere 97
SARS 156
Secreted proteins 95
Self-organizing maps (SOM) 19
Sequencing 8
Serum 78
SGD Gene Ontology Term Finder 58
Side effects 71
Signaling 8
Significance Analysis of Microarrays (SAM) 79,136
Silencing 81
Simulated annealing 44
Single linkage 18
Single nucleotide polymorphism *see* SNP
SLC29A1 80
Small round blue cell tumors (SRBCT) 103–104
SMARC4 63
SNP 8,13,72,155–156
Software 23–24,34,56–59,82,84,105–108,116–117,124,130–133,137
  AnnBuilder 137
  Array Expres 23
  Bioconductor 124,130,133
  Cloe PK 84
  COMPACT 84
  COMPARE 34
  DEREK 84
  EASE (Expression Analysis Systematic Explorer) 56
  Expression Profiler 24
  FatiGO 57
  fitPLM 130–131
  GastroPlus 84
  Gene Cluster 105–107,116
  Genesis 108,117
  GOMapper 59
  GoMiner 57
  GoSurfer 57
  Hazard Expert 84
  LeadScope 84
  MAPPFinder 57
  META 84
  MetabolExpert 84
  METEOR 84
  MultiCASE 84
  Netaffx 137
  PathArt 82
  Pathway Assist 82
  Pathways Analysis 82
  Resolver 82
  RESOURCERER 137
  Simcyp 84
SOM *see* Self-organizing maps
SOS genes 41,45,49
Sp1 153
Species 13,32–33,36,45–47,66,81,91,97–98,103,147–148,152,153,155,157
  Arabidopsis 147
  atlas 97
  comparison 97–98
  Cynomolgus Monkey (Macaca fascicularis) 91
  Drosophila melanogaster 147,153
  E. coli 147,148
  Human, Monkey, Rat 97–98
  Mycobacterium tuberculosis 13,36
  Saccharomices cerevisiae 32–33,45–47,66,81,103,147–148,152,155
  Streptococcus 13
  Yeast 152,155
  Yersinia pestis 157
  Yersinia pseudotuberculosis 157
Specificity, tissue expression 91
Splice variants 2
Spotted arrays 2
Statistical validation 17,135,136
Sterol metabolism 32
Stress response 66–68,81
Stroma 127
Structure-activity relationship (SAR) 30
Stuttgart Neural Network Simulator (SNNS) 110,117
Substance, chemical 71
Superparamagnetic clustering 82
Supervised learning 78

Supervised methods 18,22,31–34,38,102
Supervised neural networks (SNNs) 109,110
Support Vector Machines (SVM) 23,33,39,78–79,107–113
Surrogate marker 74
SVM *see* Support Vector Machines
SWI/SNP 64
SWI4 66
SWI6 66
Synapse, neural 156
Systems biology 83

T

TAA *see* Tumor-associated antigen
Tabecular bone 94–95
Target 30–32,45,97–99
  druggable 30
  genes 97–99
  identification 30–32
  validation 30–32,45
T-cell 124–125
Term enrichment analysis 56
Tes 128–129
Tibia 94–95
Tiling arrays 147–148
time series network identification (TSNI) 48
TISS (titration invariant similarity score) 36
Tissue 74,89–99,149
  brain 97
  cluster 97
  comparison 95
  expression 89–99
  expression induced by microRNA 149
  expression microRNA 149
  liver 74
  mRNA abundance 89–99
  non-coding RNA 149
  similarity 95,98
  specificity microRNA 149
Titration invariant similarity score *see* TISS
Topoisomerase type II 49
Toxic liabilities 78
Toxic response 74
Toxicant 75–76
Toxicity 71–74,76
  cellular 73–74
  testicular 76
Toxicogenomics 71–74,80
  in vitro studies 80
Toxicological response 73
Toxicology 37
Toxin 71
TP53INP1 63
TPBP1 63
Trait, complex 156
Transcript, non-polyadenylated 148
Transcription factor 12–13,40,63,97–99,152–154
  binding sites 12–13,40,63
  target 97–99
Transcription profile 2–10,125
  comparison 125
Transcriptionally functional units 147
Transcriptome 2,147
Transcript, non-coding 148–149
Transcript, prediction 148
Transforming growth factor beta-stimulated clone 22, 76
Transporters 71
Transportome 80
Tri-methylation 154
T-test 79,102,136
Tumor 12,121–130,151,153,156
  breast 12,121–130,153
  gene expression 124–129
  genome variation 156
  immunology 124
  methylation 151
  progression 121–124
Tumor-associated antigen 127–128
Two-dimensional hierarchical clustering 32,35
Type I error 136
Tyrosine kinase 122

U

Unsupervised methods 18,31–32,102
UPGMA (average linkage) 18
UPGMC (centroid linkage) 18

V

Vaccination 123–128
  prime-and-boost 124,126
  triplex 124
Vaccine 157
Vaccine, DNA plasmid 123
Vertebra 94–95
Virus 149,156–157
  animal 157
  coronavirus 156
  genus 156
  influenza 157
  microRNA 149
  SARS 156
Voltage-gated sodium channels 90
Voting algorithm 38

W

Wilcoxon test 79

Y

Yeast, budding 156

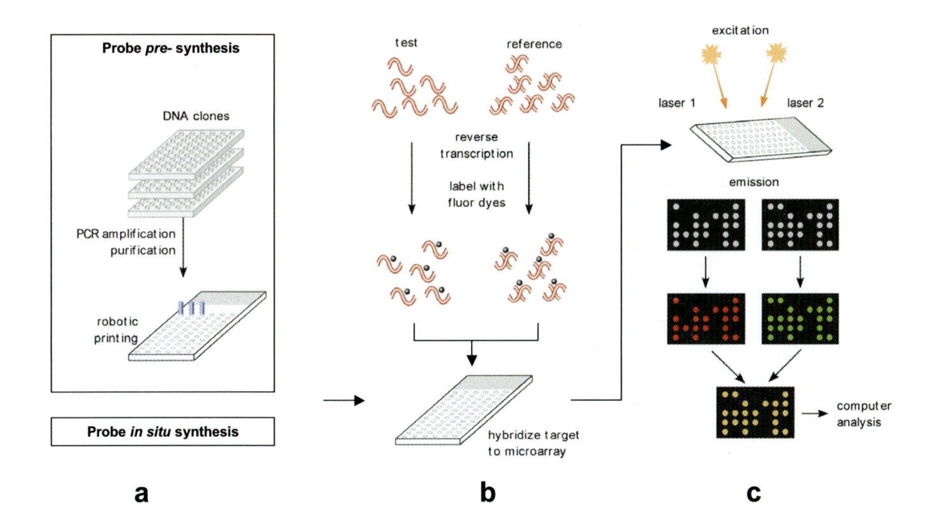

Figure 1.1 Gene expression profiling experiment on a double channel microarray. (a) Probe sequences are deposited on the chip by presynthesis or *in situ* synthesis. (b) mRNA from a test and a reference sample is extracted, reverse transcribed and amplified. Labeling molecules are incorporated during the synthesis of the amplification products. The two labeled samples are then hybridized on the chip. (c) The chip is excited at different wavelengths, one for each of the fluorophores used and the fluorescence intensities of the spots are measured. Red fluorescence of each spot measures mRNAs abundances in one sample (e.g., test) and the green fluorescence measures the same mRNAs in the other sample (e.g., reference). The two images can be merged: red spots and green spots in the combined image indicate respectively a prevalence of probe hybridization in the test sample (gene upregulation), or in the reference sample (gene downregulation). Yellow spots in the combined image indicate equally expressed genes in the test and reference samples.

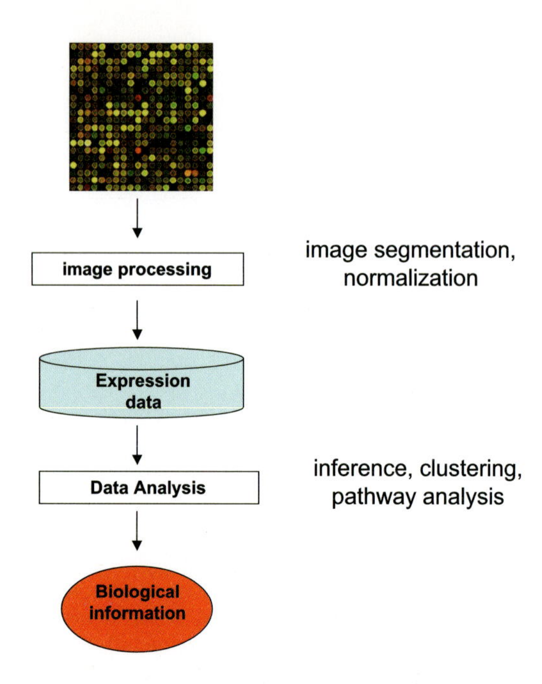

Figure 1.3 Scheme of a typical data analysis workflow. The images produced upon scanning are processed and data are normalized to correct for technical sources of variability. The resulting gene expression data can then be stored in a database and analyzed by pathway analysis, clustering and inference techniques to extract relevant biological information.

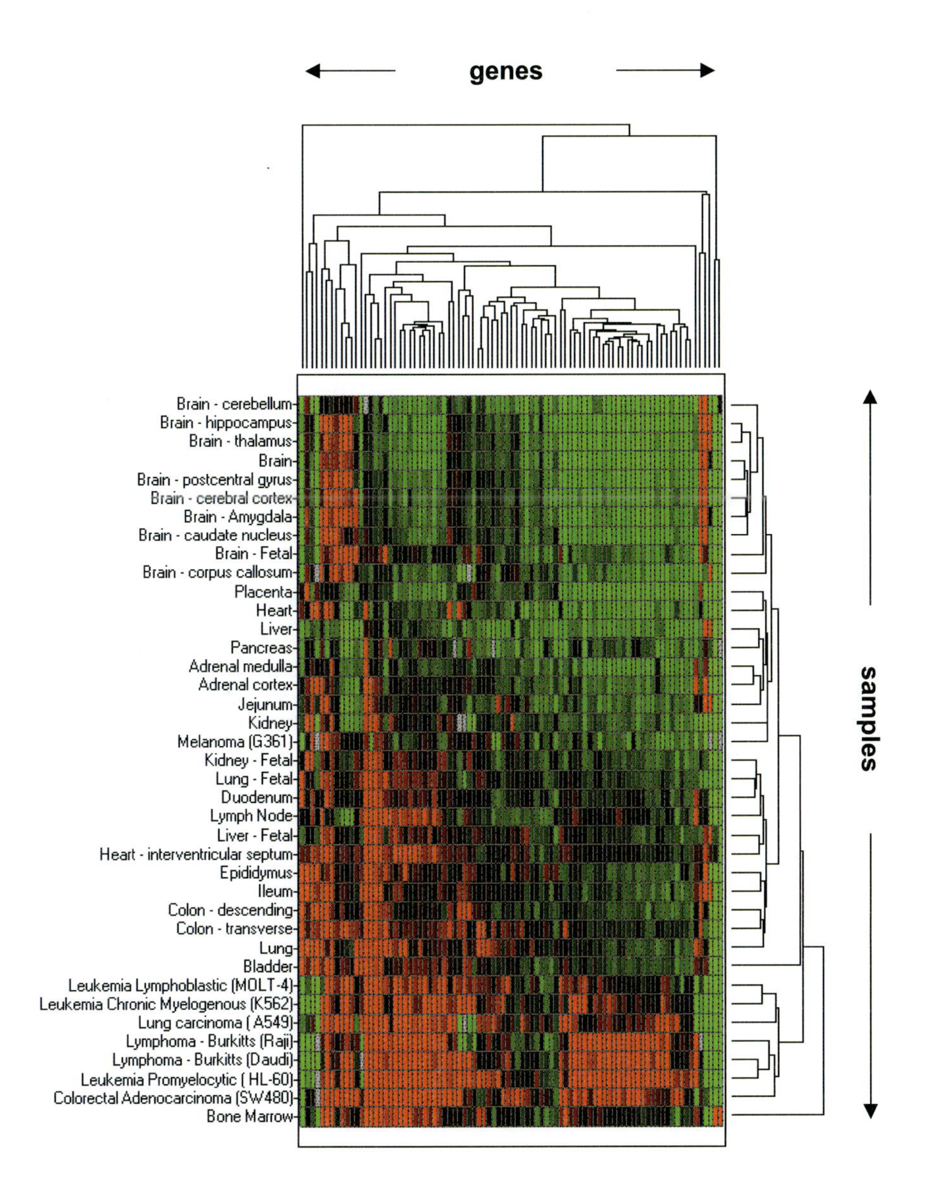

Figure 1.4 2D hierarchical clustering. Both the genes and the samples are clustered using a hierarchical clustering algorithm. The result can be represented graphically by an ordered *heatmap* diagram. Data rows (genes) and columns (samples) are reordered according to the tree defined by the hierarchical clustering algorithm. As a result, the proximity of two genes in one dimension (vertical dimension) indicates similarity in their expression pattern across the various samples assayed and the proximity of the samples represented on the orthogonal dimension (horizontal dimension) indicates general transcriptional similarities between the samples. The colors indicate the expression values of the genes: in green are usually the genes down regulated and in red the genes upregulated with respect to a baseline sample.

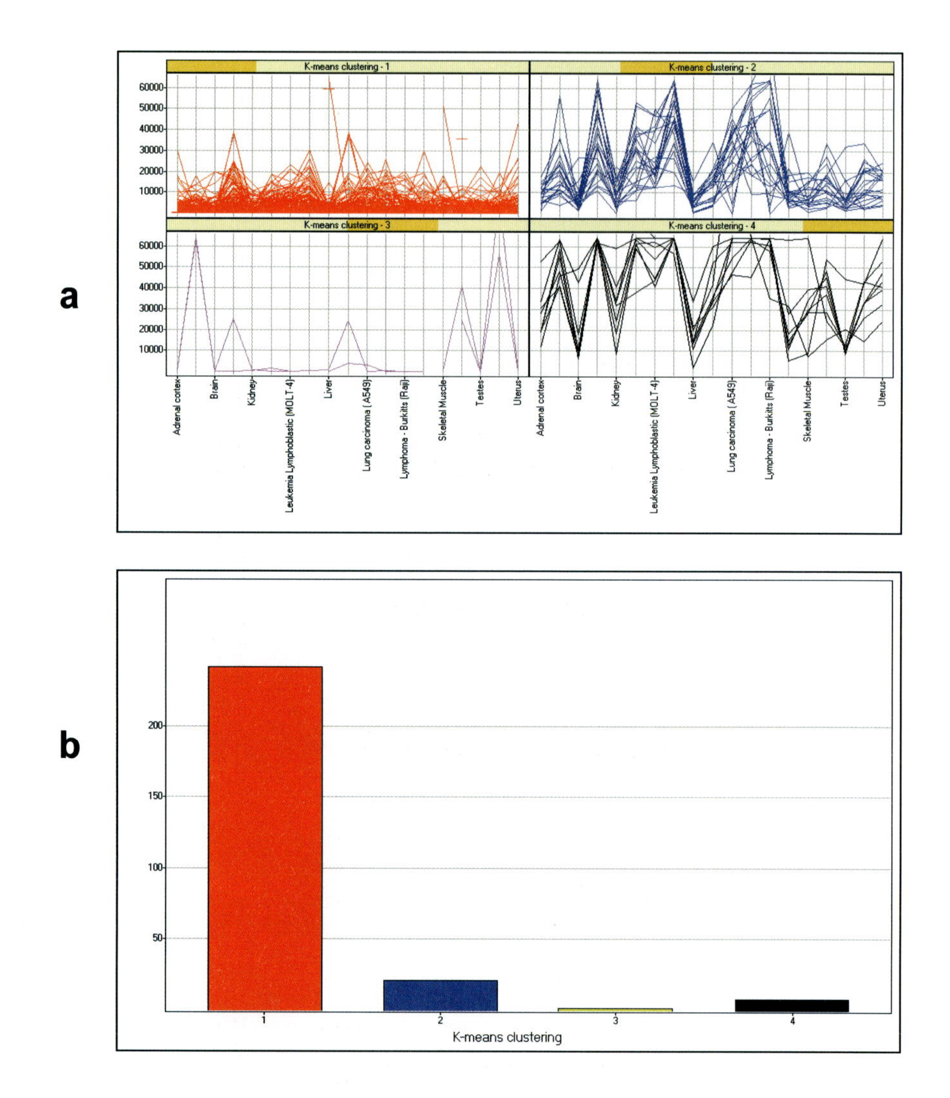

Figure 1.5 *k*-means clustering. Example of the gene clusters created applying the algorithm with $k = 4$ on an arbitrary expression dataset. (a) Four distinct gene expression cluster created by the *k*-means algorithm. Each panel shows the expression profile of the cluster genes across the samples used for profiling. (b) The number of genes falling into each different cluster can be represented by an histogram.

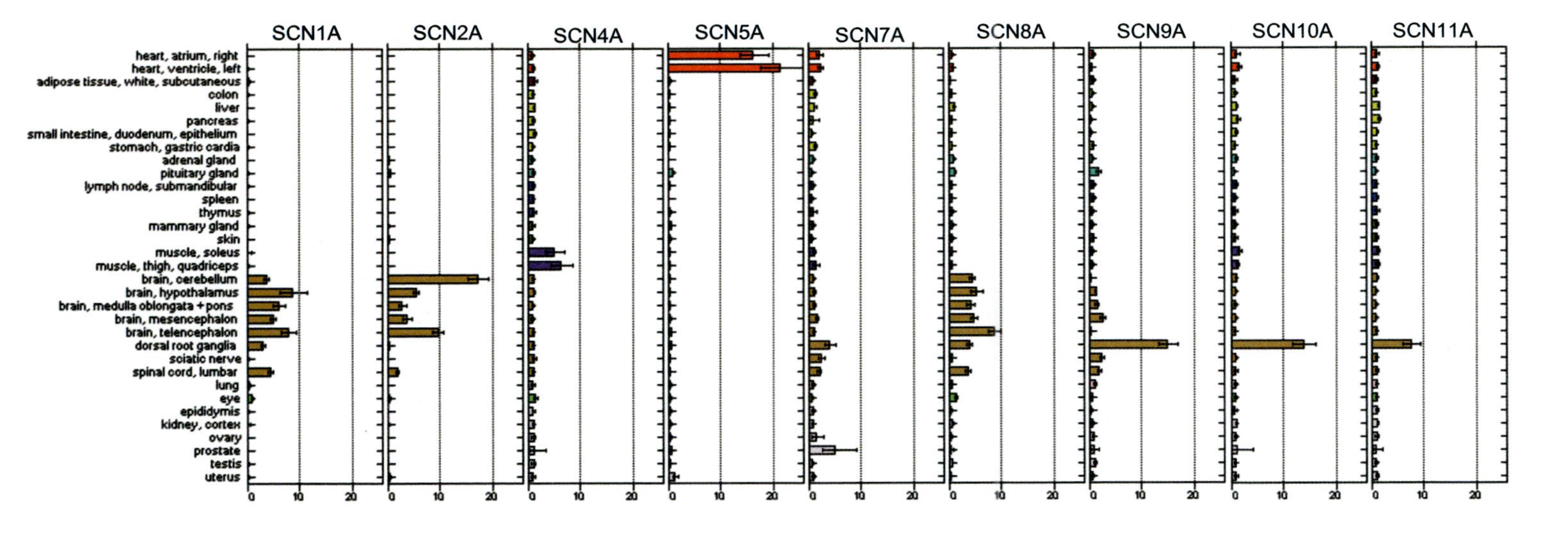

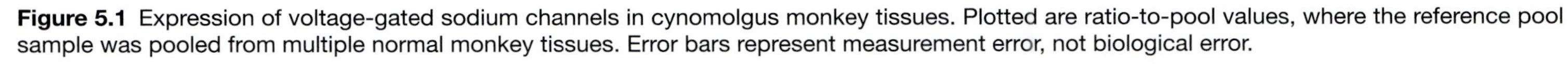

Figure 5.1 Expression of voltage-gated sodium channels in cynomolgus monkey tissues. Plotted are ratio-to-pool values, where the reference pool sample was pooled from multiple normal monkey tissues. Error bars represent measurement error, not biological error.

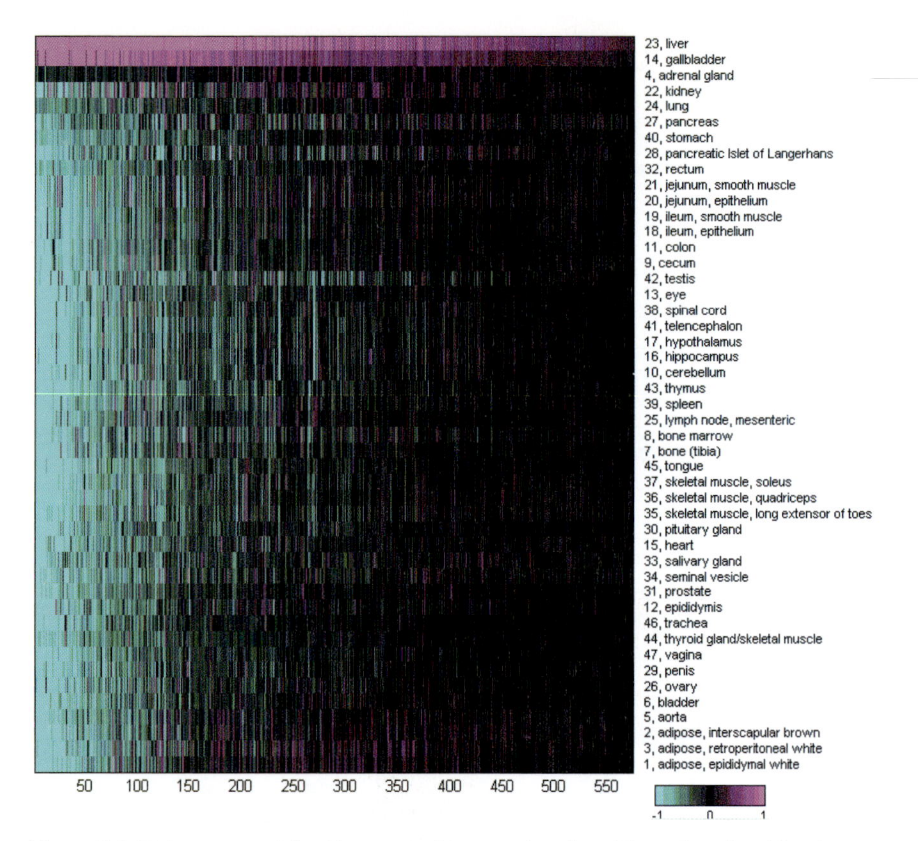

Figure 5.3 574 genes enriched in cynomolgus monkey liver. Plotted are log 10 ratio-to-pool values, where the reference pool sample was pooled from multiple normal monkey tissues. Values range from –2 to 2 and are clipped at –1 to 1 for this figure. Genes are ordered based on specificity and tissues are clustered using agglomerative hierarchical clustering.

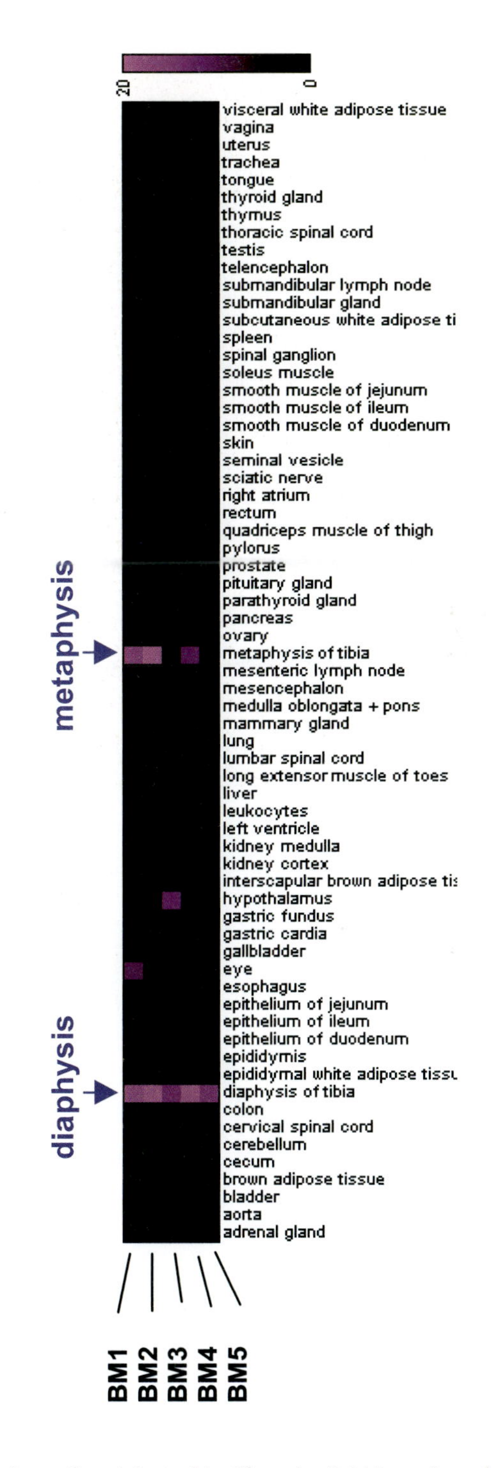

Figure 5.4 Cynomolgus monkey atlas data to identify potential biomarkers for bone growth. Plotted are log 10 ratio-to-pool values, where the reference pool sample was pooled from multiple normal monkey tissues.

A

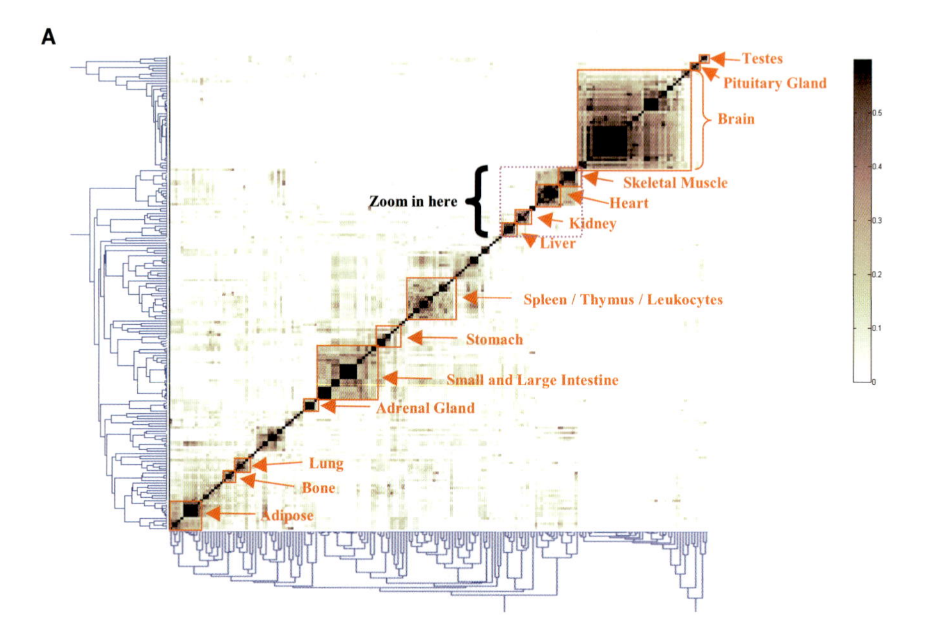

B

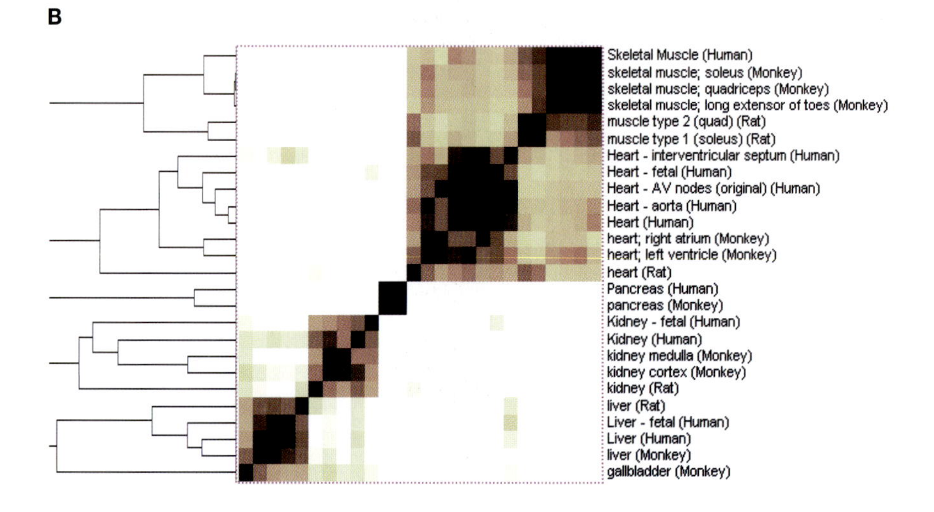

Figure 5.5 Co-clustering of human, monkey, and rat tissue profiles shows conservation of tissue-specific genes across species. 86 human, 65 monkey, and 38 rat tissue profiles are clustered using hierarchical agglomerative clustering on both axes based on the expression of 12,000 orthologous genes. Values used are ratio-to-pool, where the pool is a species-specific mixture of normal tissues. Color represents correlation, with correlation capped at 0.6. (A) All tissues examined; (B) a zoom view into a smaller region, as indicated by the dashed box.

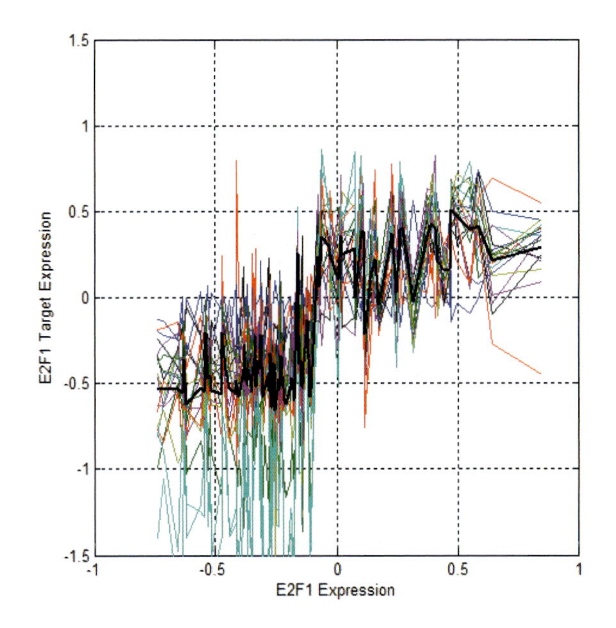

Figure 5.6 The expression of 21 targets of transcription factor E2F1 (y-axis) are plotted against the expression of transcription factor E2F1 (x-axis). The 21 E2F1 targets are determined from literature. Plotted values are log 10 ratio-to-reference pool values determined from 105 human tissues and cell lines. The reference pool is a mixture of normal human tissues. Each line represents the values for a single gene. The thick line shows the average value, across the 21 genes in each of the 105 tissues.